C0-AJQ-606

The Business Case for Web-Based Training

Dedicated to Thom, Ainsley, Morgan, and Laurel
Tammy Whalen

Dedicated to Mina, Leila, and Ramin
David Wright

For a listing of recent titles in the *Artech House Telecommunications Library*, turn to the back of this book.

The Business Case for Web-Based Training

Tammy Whalen
David Wright

Artech House
Boston • London
www.artechhouse.com

Library of Congress Cataloging-in-Publication Data
Whalen, Tammy.
The business case for Web-based training / Tammy Whalen, David Wright.
p. cm.—(Artech House telecommunications library)
Includes bibliographical references and index.
ISBN 1-58053-115-6 (alk. paper)
1. Employees—Training of—Computer network resources. 2. Employees—Training of—Computer-assisted instruction. I. Wright, David, 1947– . II. Title. III. Series.
HF5549.5.T7 W477 2000
658.3'124'0785571376—dc21 00-035575
CIP

British Library Cataloguing in Publication Data
Whalen, Tammy
The business case for web-based training.—(Artech House telecommunications library) 1. Employees—Training of—Computer network resources 2. Employees—Training of—Computer-assisted instruction 3. Employees—Training of—Computer network resources—Costs 4. Distance education—Computer network resources 5. Distance education—Computer network resources—Costs
I. Title II. Wright, David
658.3'124'0285'4678
ISBN 1-58053-115-6

Cover design by Gary Ragaglia

685 Canton Street
Norwood, MA 02062

International Standard Book Number: 1-58053-115-6
Library of Congress Catalog Card Number: 00-035575
10 9 8 7 6 5 4 3 2 1

Contents

Preface

Distance learning and the Web are two of the most rapidly developing areas in the information technology arena. The growth of the Internet is a result of the following.

- Continually increasing multimedia capabilities;
- Broad geographic accessibility;
- The number and sophistication of users.

Distance learning is a growth area from the viewpoints of the following entities.

- Corporations and nonprofit organizations concerned with rapidly training large numbers of employees at reasonable cost with minimal disruption of work;

- Public educational institutions, which see Web-based learning as a cost-saving measure in a time of reduced budgets;
- Producers of software for Web-based course development and delivery;
- Multimedia production and instructional design companies that package course content and offer consulting services to consumers of Web-based courses;
- Telecommunications equipment vendors and public carriers, which see distance learning as a major application for the multimedia capabilities of the Internet.

The majority of books currently on the market discuss the "how to" of distance learning—the procedures necessary to put teaching materials on the Internet—while other works discuss the many technologies used for distance learning without providing a clear, in-depth focus on the Web. This book evaluates the relative merits and uses of Web-based distance learning and addresses the management issues—cost benefit, marketing, pricing of Web-based courses, and business process reengineering—that are faced by organizations in their implementation of it. The book's management and economic focus will help answer the question of why organizations should undertake Web-based training. While the book is written primarily from the perspective of the corporation, both nonprofit and public sector institutions will benefit, as the issues faced are similar.

Some of the intended audiences for this book are the following.

- Training department personnel in corporations and nonprofit organizations concerned with timely and cost-effective delivery of training to employees;
- Administrators in public educational institutions concerned with management issues associated with distance learning over the Internet;

- Teachers at secondary and postsecondary institutions concerned with understanding the implications of providing education over the Internet;
- Software applications developers concerned with building features into their products that meet organizational requirements for distance learning.

This book is divided into six chapters, each illustrating a key concept related to the economic and business impacts of implementing Web-based training in corporations. Chapter 1 presents an overview of the Web-based training industry and includes examples of how corporations are using Web-based training today. In addition, Chapter 1 introduces the costs and benefits of Web-based training, including online training management, and analyzes Web-based training from a marketing point of view, with a discussion of the "4 P's": product, place, price, and promotion. Finally, it presents key factors in the implementation of Web-based training, including business process reengineering (BPR) considerations as well as the potential barriers to the adoption of Web-based training.

Chapter 2 hypothesizes that there are several key design elements that must be considered in costing Web-based training projects. Educators, trainers, and businesspeople need to be able to evaluate the cost-effectiveness of Web-based training to make informed decisions about the extent that their organizations should use this new media. The relative importance of these elements is examined using a case study approach. The case study provides a detailed cost-benefit analysis, including the break-even number of students required to recover Web-based course development costs and the return on investment over a five-year period. The methodology used in this case study can be used in future cost-benefit studies of Web-based training.

Chapter 3 further sets the context for Web-based training by comparing the costs of these two major modes of delivery for distance learning. Examples of uses of these technologies are given for academic institutions, government, corporate training applications, and K–12 schools. Chapter 3 also discusses the importance of collaborative learning, as well as how these technologies are evolving into hybrids that blur the distinctions between videoconferencing and Web-based communications.

Chapter 4 discusses the value to companies of using a Web-based system for competency and training management, using the case of Bell Canada as an example of how companies are implementing these tools today. The Web has had a major impact on how corporate training departments manage employee training. The evolution of computers and networks allows companies to implement a precise customer-focused approach. Through the use of competency and training management systems such as the Sytéme Intégré de Gestion d'Apprentissage en Ligne (SIGAL) system used by Bell Canada, organizational training plans can be efficiently communicated throughout the organization, training needs can be linked to the performance evaluations of individual employees, and online training materials can be conveniently delivered to employees at their desktops.

Chapter 5 examines the factors that influence the pricing of Web-based courses for employee training and discusses pricing models currently used by vendors competing in the marketplace, as well as the market forces that affect the level of competition, such as the availability of substitutes. By evaluating several pricing scenarios, we consider the implications of various pricing models with respect to course development costs, profit margin, payback period, net present value, and return on investment. Using the case study of Web-based training at Bell Nexxia, we examine the appropriate price to charge internal business units for Web-based courses offered to employees and discuss whether the prices to external customers should be the same.

Chapter 6 is a case study that examines the use of Web-based training at Bell Canada in the context of BPR. This case study describes changes in the organization that result from reengineering, including the impact Web-based learning has on training plans, student needs assessments, the ability to provide a specialized curricula, training students and instructors in using new technologies, and establishing a principle of continuous improvement. Alternative ways of achieving project objectives are presented, along with organizational impact, technology alternatives, and cost benefits. The study defines the processes that are required to deliver Web-based training, the value to the internal and external business practices of the organization, and the costs for each process. Chapter 6 identifies the wider applications of this case study, which will be of interest to those in organizations moving from classroom-delivered training to distance delivery.

This book was funded in part by Industry Canada's Network of Centres of Excellence in Tele-Learning (TL-NCE).

Introduction

The information age has brought many changes to the workforce and heightened the need for workplace training. Competitive advantage is closely linked to the ability of employees to respond quickly to change. Today's "knowledge worker" has an ongoing need for both new information and updated skills to respond to challenging and constantly changing job demands. Rapid response requires employees to know how to perform new job functions, and Web-based training plays an important role by allowing the delivery of courses and training modules in a timely, convenient, easily updated, and cost-effective manner. Among the top 10 trends in training [1] are the following.

- Increased requirements for job-related skills;
- Increased focus on performance improvement and human performance management;
- Advanced technology and revolutionized delivery systems.

Technology-based distance learning has its origins in computer-based training (CBT). Large companies with a high need for alternative methods of training employees were the first to adopt CBT. IBM, for example, started using CBT to train employees in the late 1960s. Computer technicians, typically working almost exclusively at client sites maintaining IBM mainframe computers, took training related to their jobs using the computers they serviced. The use of distance learning has become more widespread as computers have become more powerful and high-bandwidth networks extend their reach throughout the corporation. Today, many companies are implementing Web-based training that is delivered to desktops, computers in employees' homes, and laptops at any location through corporate intranets, extranets, or the Internet.

Benefits of Web-based training

Using Web-based courses for employee training has both qualitative and quantitative benefits. Qualitative benefits include the following.

- Convenience to employees in terms of training location and use of familiar technologies;
- Access to expert instructors regardless of geographic location;
- Added value to the learning experience through the interactivity of technology-assisted instruction;
- Increased employee access to training due to the elimination of scheduling restrictions and the reduced costs of training delivery.

Apart from cost savings, calculated as a return on investment, quantitative benefits related to learning efficiency and retention [2] include the following.

- Faster (by 60%) learning curve;
- Higher (by 25–60%) content retention;
- Greater (by 56%) learning gains;

- Better (by 50–60%) consistency of learning;
- Faster (by 38–70%) training comprehension.

Evaluation of Web-based training

Throughout this book, we discuss ways that organizations can evaluate the impact that training has on the organization, from both a cost-benefit and a business perspective. The four-level Kirkpatrick system, outlined in Table I.1, is an industry standard for training evaluation [3].

Training evaluation at the "results" level becomes especially important in justifying training expenditures. However, more than 80% of U.S. organizations still evaluate training based on feedback from trainees (level 1), and fewer than 50% evaluate training in terms of business results that are represented by the higher levels of the Kirkpatrick model [4]. A survey of Canadian companies found similar trends [5], with the following percentages of the courses offered within training departments evaluated at each of the Kirkpatrick levels:

- Reaction: 84%;
- Learning: 42%;
- Transfer of learning: 23%;
- Results: 16%.

Davenport [6] argues that learning technologies are forcing information technology and human resources departments into the same niche, along with electronic performance-support systems (EPSS) and

Table I.1

Kirkpatrick's Four Levels of Training Evaluation

Level 1	Reaction: How well participants liked the training
Level 2	Learning: Principles, facts, and techniques that were understood and absorbed by the participants
Level 3	Transfer of learning: Transfer of training skills to knowledge on the job
Level 4	Results: Impact of training on the organization

Adapted from [3].

knowledge management. Companies such as GES Exposition Services in Las Vegas report that customer calls to their company help desk are later analyzed to determine trends in requests for assistance that should be addressed through online training courses targeted at customers. The company feels that this approach is an effective way to tie return on investment for training costs to a demonstrated business need [7]. A survey of American companies indicated that human resources and development (HRD) professionals in many organizations are already responsible for identifying, implementing, and maintaining HR systems for use in training, as well as designing and maintaining training content itself [8]. Web-based training that makes specific training available when needed, right at employees' desktops and integrated with Web-based training management systems, will be one of the primary links between the information technology, human resources, and training departments.

Does training really pay off? Many companies believe that it does. For example, the American Bankers Insurance Group reported that extensive training resulted in a 30% rise in sales in 1994 and an average 20% growth in sales for each of the five previous years [9]. While figures such as these are encouraging, hard numbers are often lacking when it comes to measuring the success of a training investment. As Peter Senge points out, organizations never get to see what would have happened if an investment in learning wasn't made [10]. Evidence of the benefits of organizational learning is generally anecdotal. For example, employees at one American company have cited the availability of training opportunities and ability to take on more challenging job tasks as the main reasons they choose to stay with the firm despite attractive offers from other companies [11].

Anecdotal evidence of the benefits of training can be useful, but a business case for a new training investment has to rely on more than assumptions. The most straightforward type of business case for the use of technology-enabled learning is that which this book describes, the comparison of the use of technology to traditional alternatives. Through training management systems, measuring the impact that training has on the organization becomes more feasible, even though training evaluation is still closer to an art than a science. Making a business case for Web-based training delivery is a critical step in ensuring that the mode of training delivery maximizes the money spent on training. While this book does not address the question of how much an organization should invest in training, it can provide a framework for deciding how those dollars should be spent.

References

[1] Bassi, Laurie J., Scott Cheney, and Mark Van Buren, "Training Industry Trends 1997," *Training and Development*, Vol. 51, No. 11, November 1997, pp. 46–59.

[2] Ciancarelli, Agatha, "Online Training May Become Preferred Method," *Purchasing*, Vol. 125, No. 9, December 10, 1998, pp. S35–S36.

[3] Kirkpatrick, Donald L., "Techniques for Evaluating Training Programs," *Training and Development*, Vol. 33, No. 6, June 1979, p. 78.

[4] National Alliance of Business, Inc., "Company Training and Education: Who Does It, Who Gets It and Does It Pay Off?" *Workforce Economics*, Vol. 3, No. 2, June 1997, pp. 3–7.

[5] Benson, George, "Is Training Different Across the Border?" *Training and Development*, Vol. 51, No. 10, October 1997, pp. 57–58.

[6] Davenport, Tom, "HR and IT in Wedded Bliss," *CIO Magazine* (http://www.cio.com), May 1999, pp. 34–37.

[7] Lakewood Publications Inc., "Mating H.R. and I.T. at the Help Desk," *Online Learning News*, Vol. 2, No. 7, May 18, 1999.

[8] Bassi, Laurie J., Scott Cheney, and Mark Van Buren, "Training Industry Trends 1997," *Training and Development*, Vol. 51, No. 11, November 1997, pp. 46–59.

[9] Cohen, Andy, "Training for Success," *Sales and Marketing Management*, Vol. 147, No. 7, March 1995, p. 27.

[10] Lakewood Publications Inc., "Why Organizations Still Aren't Learning," *Training*, Vol. 36, No. 9, September 1999, pp. 40–49.

[11] Dobbs, Kevin, "Winning the Retention Game," *Training*, Vol. 36, No. 9, September 1999, pp. 50–56.

1

The Market for Web-Based Training

1.1 Market demand for Web-based courses

Company expenditures for employee training are increasing. In 1997 U.S. companies spent between $55 billion and $60 billion on all types of education and training, an increase of 18% over adjusted 1985 spending [1]. These figures represent 2.1% of payroll for U.S. companies and 1.6% of payroll for Canadian companies, although a possible explanation for this discrepancy is that the U.S. sample contained a greater proportion of large firms [2]. Large companies are about 1.4 times as likely to offer employees training as smaller companies, and the use of external training providers such as equipment vendors, private consultants, industry associations, and technical and community colleges is increasing. In North America training is most often given to workers with higher levels of education who are less than 55 years of age and working in managerial and professional occupations in medium and large companies [3].

As for which occupations receive this training, salespeople, followed by management professionals, are most likely to be trained. Although production workers receive the most overall hours of training, this is due to the fact that production workers outnumber salespeople and management professionals in the workforce by almost 2 to 1. While the majority of employee training is still offered in a classroom setting, more and more companies are offering Web-based courses as an alternative to the classroom, or to supplement a classroom course. Computer skills was the most common topic taught via distance learning, making up 57% of the distance courses taught in the organizations surveyed. Mobility among high-tech workers and rapidly changing technologies contribute greatly to the need for training related to information technology (IT). For these reasons, some large U.S. systems integrators are experiencing close to a 100% annual increase in training expenses [4].

The demand for Web-based training comes primarily from current consumers of traditional classroom training, such as medium and large companies and government. Universities and colleges also use the Web for the delivery of some of the courses they offer, particularly to mature students. Training, compared with the broader category of education, is characterized by an emphasis on job-related skills. Web-based training may be used as a replacement for classroom courses but is more commonly used as a supplement to classroom delivery.

The Masie Center of Saratoga Springs, New York, estimated that the 1999 market for online learning in North America was approximately $600 million [5]. This marks a substantial increase over previous years. In 1996 expenditures on Web-based training by U.S. companies amounted to $100 million [6]. While 81% of employee training was delivered live in the classroom in 1987, only 70% of training was classroom-based in 1998 [7]. In other words, from 1987 to 1998, technology-based training experienced an average annual growth rate of 1% per year in the United States. The situation is similar in Canada, where 70% of companies surveyed in 1997 reported that they used some form of technology-based training [8].

Most online training today is delivered on CD-ROM disks or over corporate intranets, and 42% of online learning includes interaction with an instructor or other students. Intranet delivery is especially popular in the United States. In 1997, 70% of U.S. companies used intranets for training, compared with 6% of companies in Britain [9]. In Canada, 11% of companies used intranet-based course delivery and 10% used the Internet for training delivery [10].

A survey of Fortune 1000 companies by the Masie Center found that training departments that are early adopters of online learning have senior executive support but tend to encounter resistance from IT personnel as well as some managers and learners [11]. The survey also determined the following.

- Most (89%) of the companies were implementing online training delivery, primarily for IT applications and to train sales staff working from remote locations while on the road.
- A majority (63%) of the respondents claimed that their corporate culture resists online training delivery and that business process issues—for example, the pricing of online learning courses—pose their largest obstacles to implementing online training programs.
- A majority (72%) of learning-technology professionals claimed that they would like to see learning products that are not yet in the marketplace, such as software with more precreated content, more simulations, and easy-to-use course authoring templates.

Some critics argue that functionality must necessarily take precedence over ease of use in online training products. David Liddle predicts that purchasers of learning products will look first for features that respond to their needs and believes that trainers will accept complexity in learning products that are fully functional [12].

This chapter profiles the Web-based training industry and provides specific examples of its use in corporations today. Web-based training is analyzed from a marketing point of view: product, place, price, and promotion. Finally, key factors in the implementation of Web-based training are presented, including BPR considerations, more fully explored in Chapter 6, as well as the potential barriers to the adoption of Web-based training.

1.1.1 Market segmentation

Market segmentation refers to the analysis of potential buyers of a product. Key questions that help determine the characteristics of potential consumers are: Who will buy this product? What are the needs of those buyers? What are their key purchase criteria? The market segments that form the primary consumers of Web-based training products are the following.

- Medium and large companies and nonprofit institutions;
- Government;
- Training institutions.

Some of the indicators that an organization has a high need for Web-based training include the following.

- Many training activities are required by a particular employee segment or for all workers on a particular topic.
- Training content changes rapidly and must be updated as cost-effectively as possible.
- Employees are currently equipped with computers and network connections for their daily tasks. (Web-based training would suit their normal mode of working as well as the company culture.)

Surveys of Canadian companies have found that the oil and gas industry has the highest training expenditure per employee, followed by the finance, insurance, and real estate industries, and then the mining industry. The education, health, and manufacturing industries tend to spend the lowest amounts per employee, although manufacturing, as well as the oil and gas industry and the wholesale and retail trades, is predicted to experience the largest increase over the next few years [13].

1.1.1.1 Demand for IT training and professional certification

The community of interest for Web-based learning is far broader than the traditional education and training communities, which are typically comprised of younger full-time students and older full-time workers. For example, SeniorTechs is a job referral service for 15,000 unemployed "vintage techies" (i.e., over 35 years old) seeking work. SeniorTechs intends to offer members online training courses to upgrade their skills in the latest programming languages and in related skills [14]. Within the IT mainstream, Sears is an example of a company that has embraced online learning by giving its IT professionals access to approximately 800 Web-based courses [15].

Consumer demand for IT training has prompted vendors such as Microsoft and Novell to offer Internet seminar and certification courses via the Web [16]. Professional associations such as the International Webmasters Association and the Association of Internet Professionals are also developing certification programs [17]. A survey of both IT and non-IT professionals in December 1998 by technology training company USWeb Learning found that 72% of the 2,400 professionals surveyed believed that they would benefit from Internet training and certification [18].

Even the medical profession, which traditionally values "hands-on" experience, is finding a way to use the Web for continuous learning. Cornell University is using a video-streaming-on-demand technology to deliver lectures, graphics, and photographic materials on the latest medical developments to doctors' desktop computers [19]. The U.S. military is also adopting Web-based training for both medical and IT professionals [20]. Several sites allow medical specialists to take continuing medical education courses via the Web to retain their credentials [21]. These sites include the University of Florida (www.medinfo.ufl.edu/cme/cme/modules.html), Virtual Patient at Marshall University (medicus.marshall.edu/mainmenu.htm), NetCME (netcme.mdanderson.org), and Clinical Care Options in HIV (www.healthcg.com/hiv).

Web-based professional certification is also being done in nursing and the financial industry [22]. SpringHouse (www.springnet.com) and Health Interactive (www.rnceus.com) are both Web-based learning sites where nurses can take accredited training and state-approved testing for recertification. After nurses have taken the online courses and passed the necessary tests, their certifications are e-mailed to them. The service has been used by thousands of nurses with great success. Similarly, Web-based continuing education courses for bankers, insurance professionals, certified public accountants, and real estate agents are offered by IBT Financial Inc. (www.ibtfinancial.com). The company offers Web-based courses and testing, e-mails candidates a notice of certification after successful completion, and keeps a database of student records for verification by licensing boards.

Many industries have begun to use the Web to make training directly available to their own communities of interest as well as help members locate training from other sources. For example, in 1999 the American Institute of Chartered Public Accountants began to explore the use of Web-based training by making a half-dozen credit courses available on its Web site [23]. Many other diverse industries are also using the Web to train workers. Fleet managers in the trucking industry are able to check the Web site www.truklink.com to discover where training is available on many working-related subjects [24]. Steelcase Inc.'s employees and 450 dealers in North America can use its Web site to take training courses and obtain career development information, which makes training more accessible as well as saves the company $70,000 each year in training administration and materials costs [25]. Budget Group Inc.'s online solution allows employees to learn about the point-of-sale system used by the company [26].

Professional training companies are also incorporating Web-based training into their curriculum to meet the demand for certification, although start-ups appear to be further ahead when it comes to making Web-based courses available to customers. 7thStreet.com and Cytation.com offer more than 150 IT courses through America Online (AOL) [27]. Similarly, Learning.Net offers continuing education and certification courses through its Web site. Some examples of Web sites that anyone can visit to research the availability of online training courses are the American Society for Training and Development Seminar Agent (www.astd.org/virtual_community/seminar_agent), Wayne's Comprehensive Computer Professional Certification

Resource (www.diac.com/~wlin/cpert.html), the IT Training Buyer's Guide (www.ittrain.com/guide/guide-index-product.html), the Training Registry Inc. (www.tregistry.com), and Education New Brunswick (http://teleeducation.nb.ca/).

1.1.1.2 Link between higher education and training

Another area well represented in the Web-based course market is accredited university courses. A large number of universities now offer online programs, some without residency requirements. For example, Athabasca University, the University of Phoenix, National University in San Diego, and the University of Maryland at Baltimore are just a few universities with extensive degree programs offered via the Internet. About 250 universities worldwide, including brand-name institutions such as Duke, Stanford, and MIT, have online course offerings. The worldwide virtual student body was estimated to be 750,000 in 1999, and it is predicted to double by 2004 [28]. Other analysts estimate that enrollments will total 2.2 million students by 2002 [29]. Perhaps these predictions are not surprising considering the trend toward rising university enrollments, along with increasing numbers of full-time students who are also in the workforce. One of the barriers to pursuing further education includes the inability of students to schedule classes around work and other responsibilities, and self-paced courses that are delivered over the Web are a practical alternative to classroom courses [30].

A study of 61 universities and colleges in the United States and Canada [31] determined the following.

- Most (93%) use e-mail as their primary delivery medium.
- Most (73% to 87%) reuse classroom course materials rather than create new courses for the Web.
- Increasingly, sessional lecturers who have been hired on a temporary basis to teach one or more courses teach Web-based courses.
- Salaries account for the highest percentage of the costs of distance-learning programs.

- Most (87%) of the programs operate at a profit, and 13% realize a profit of greater than 50%.

One example of the types of programs offered by universities and colleges via the Web today is represented by the programs in quality assurance from California State at Dominguez Hills (www.csudh.edu/msqahome.htm), Aurora University in Illinois (www.aurora.edu/qsm/index.html), and Roane State Community College in Harriman, Tennessee (www2.rscc.cctn.us/~clauson_jr). Countries outside North America and Europe are also interested in creating "virtual universities" in which classes are held over the Internet rather than in a traditional classroom. For example, Mexico's Instituto Tecnològico y de Estudios Superiore de Monterrey offers virtual classes to approximately 70,000 students from 27 locations across the country [32].

Some management experts are even predicting that the dominance of traditional "bricks-and-mortar" universities in providing education will be a thing of the past. In an interview in *Fortune* magazine (http://www.pathfinder.com/fortune/1998/980928/dru.html), well-known management theorist Peter Drucker predicted that in the future, online educational resources will surpass universities in importance and even threaten the survival of universities [33]. A new virtual university model is emerging, led by Western Governors University (WGU) in the United States. WGU offers its own degree programs to students studying at a distance, with the innovation of using Web-based courses authored at other universities [34].

Although initial enrollments were lower than predicted—100 students enrolled in 1999 rather than the 3,000 who were expected—the states in support of the university are optimistic about its future [35]. Eighteen states participate in WGU, each initially contributing $100,000 for the university's establishment. Other sources of start-up funding have come from companies and foundations. In the spring of 1999 WGU also requested a further $8 million from the U.S. Congress to develop curriculum, hire academic advisers, and construct local WGU centers where students can access online classes or take tests [36]. New funding from AT&T, a partnership with the Open

University in the U.K., and support from U.S. Vice President Al Gore have all strengthened WGU and confirmed public support for both the university and the role of distance learning in the future of universities [37].

Closer ties between universities and corporate training departments may be an indication that Drucker is correct in his assertion that traditional models of education are changing. An innovation in the training field is the partnership between corporate training departments and universities, which allows employees to take courses at work to earn credits toward a university degree program [38]. Columbia is one university with a clearly stated goal of making its course materials available for a fee over the Web, which it does through its business unit Morningside Ventures. Employees of corporations are one of the primary markets for these courses [39].

The public image of corporate training departments is undergoing a shift toward closer alignment with the values of traditional educational institutions. One trend is to rename the company training department a "corporate university" that offers a wide variety of training and educational opportunities, often in partnership with academic institutions. For example, the Siemens Business Communications Systems Inc.'s Virtual University offers online classes, with a live instructor, at employees' desktops. Classes are broken into short modules, and students are assigned "homework" that is integrated into their regular work. One of the benefits has been a saving of $800,000 in the first year [40]. A survey of Canadian companies found that 40% were associated with a training school, institute, or corporate university [41]. There are currently approximately 1,600 corporate universities in the United States, with 40% of the *Fortune 500* companies represented in that number [42]. Improving employee skills and corporate culture are the two main goals of these organizations, but some corporate universities are contributing to their companies' bottom lines by offering their courses to the public.

Some brokerage houses have moved a substantial portion of their business activities to the Web. Charles Schwab and Excite have joined forces to offer visitors to their sites educational information on investments [43]. Their intent is to encourage visitors to spend more time at the Web sites. Investment professionals such as stockbrokers and dealers can take mandatory continuing education required by the Securities and Exchange Commission through the site [44].

Apart from workplace training, online learning is also moving into areas that do not traditionally involve the use of computers. One such example is a product from Newfoundland company Full Tilt called the On-line Music Conservator, which allows students to take music lessons over the Internet [45].

1.1.1.3 Relationship of Web-based learning to e-commerce

Companies that want to conduct commercial activities on the Internet, including the sale of Web-based training, have to overcome significant challenges in getting customer attention and maintaining customer relationships. To assist companies with these tasks, a new type of service provider, called a metamediary, acts as an agent to provide a single point of contact between customers and suppliers, primarily by providing information and communications through Web portals [46]. For example, metamediaries working in the auto industry, such as Edmund's (www.edmunds.com), provide customers with information such as data on new and used auto pricing, dealer costs and holdbacks, auto reliability, buying advice, and auto reviews. When users visit the Edmund's Web site, the site flows visitors to the Web sites of partners in the distribution channel.

In the Web-based learning industry, online virtual communities such as "Virtual U's" bring together people with product and service needs that go beyond the purchase of courseware. A Web portal called CollegeBytes.com created by CommonPlaces (www.commonplaces.com) is a single Internet hub for college students, who may chat with professors, log on to online lectures, buy textbooks and movie tickets, register for classes, and build personal Web pages. Use of the site is free to students, and revenue is anticipated from electronic commerce-based transaction fees charged to participating colleges and universities [47].

A similar portal initiative exists in the Campus Pipeline in Salt Lake City [48]. The Campus Pipeline is a service available to universities across the United States that allows the universities to provide students with access to e-mail services, online registration, financial aid guidance, and classroom chat sessions. Although services are free to both the participating universities and students in return for member use, the Campus Pipeline requests permission from every student enrolled at the participating universities for access to their student records. The Campus Pipeline then sells that information, such as student major, grade point average, and hometown, to advertisers. A Web interface tool that universities can use to create their own portal is Jenzabar [49]. Apart from providing convenience to students, universities may themselves be able to reduce administration costs by allowing students to serve themselves when they want to obtain information or register and pay for courses.

1.2 Web-based training industry profile

1.2.1 Scope of the industry

The scope of the Web-based training industry is vast, encompassing a wide variety of hardware, software, and service suppliers for course production, delivery, communications, and management, as well as consulting and user support. The Web-based training industry has its origins in the software products and services industry and the commercial education and training industry. These products and services serve three primary markets:

1. Corporate training;
2. Postsecondary education;
3. Elementary to secondary school (K–12).

Suppliers of Web-based training are primarily small- and medium-size companies operating in the United States, and to a lesser extent in Canada, Europe, and other countries around the world.

1.2.2 Industry structure

Some of the niche markets that have developed in the Web-based training industry include multimedia course development, authoring and delivery tools, competency and training management systems, and systems integration. The industry is comprised of six product and service components:

1. Network infrastructure;
2. Data transport and data storage equipment;
3. Course development, delivery, communications and collaboration, and training management systems;
4. Course content;
5. Course hosting and customer care services;
6. Professional consulting services.

1.2.2.1 Providers of network infrastructure

Network infrastructure is provided by network service providers that own or lease the wires, cables, and wireless technologies that form the communications routes for training information. Telecommunications companies such as AT&T, MCI Worldcom, Sprint, Ameritech, and regional telephone companies in the United States, as well as Bell Canada, Atlantico, SaskTel, Manitoba Tel, and Telus (along with some of the U.S. competitors listed above) in Canada, and many others around the globe, all provide network infrastructure. Also competing in this area are cable companies such as Jones Cable and Rogers Communications. These companies provide network services for many types of applications and are an important component of end-to-end business solutions for Web-based training.

1.2.2.2 Providers of data transport and data storage equipment

Delivery of course material benefits from the continuous improvements being made in client-server technology, Internet protocol (IP) connectivity, and increased network capacity. Data transport equipment includes hubs, switches, and routers that transport training information across a network infrastructure, and data storage equipment such as client-server technologies. Cisco, Nortel, Bay Networks, Newbridge, Lucent, Sun Microsystems, Silicon Graphics, IBM, and Microsoft are some of the manufacturers of data transport and/or data processing and storage equipment. Like network service providers, equipment manufacturers provide devices for many business applications, including Web-based training.

1.2.2.3 Providers of course development, delivery, and management systems

Course development, delivery, and management systems are provided by software development companies and used for course authoring, delivery, communications and collaboration, and course management. MacNeil [50] identifies the elements of a learning system as the following.

- A graphical user interface (GUI);
- A messaging system that can include e-mail, fax, or voice messaging;
- Synchronous or asynchronous conferencing using text, images, audiographics, or videoconferencing;
- A collaboration system for file and application sharing;
- Internet client tools such as file transfer protocol (FTP) and browser links to public databases and Internet sites;
- Access to workstation applications such as word processors and spreadsheets;
- Session management, evaluation tools, and indexing systems.

1.2.2.4 Providers of course content

Course content consists of text, graphics, audio, or video elements driven by an instructional design. There are four major groups of companies providing course content and delivery to the Web-based training industry:

1. Companies providing specialized training services (e.g., Learning Tree, EDS, and freelance consultants);
2. Businesses in other industrial sectors that offer training services (e.g., IBM, Microsoft, Novell, professional accounting organizations, and telecommunications companies);
3. Private-sector training institutions (e.g., De Vry, Toronto School of Business, and Development Dimensions International);
4. Commercial training activities of public education institutions (e.g., executive MBA programs).

A list of course producers in *Training* [51] indicates that there is a larger number of vendors for IT training than for other types of courses such as soft skills (i.e., "people" skills) training, general interest, or industry-specific courses. For example, DigitalThink Inc. is one of the many companies that offer a range of online training options from live and text messaging-based online tutorials and group discussions to formal courses on many current technology topics [52].

The Masie Center estimates that there are approximately 500 products, systems, and services competing to provide online training to organizations and individual learners [53]. By comparison, there are approximately 5,300 for-profit firms operating in the entire education and training sector in the United States [54]. Resellers of IT products and services are also starting to use the Web for product training, with 15% of resellers reporting in 1998 that they use the Web to deliver training to customers [55].

Is the content of Web-based courses different from that of classroom courses? The large number of IT-related courses available for Web-based delivery may lead to the conclusion that course vendors and training departments believe that a technology-based delivery medium is particularly appropriate for courses on technology-related topics. However, a survey of organizations worldwide concluded that technology skills courses are the most common of all types of training, including classroom delivery [56]. Of the organizations surveyed, 63% offered technical skills training to most employees while less than 30%

offered career or specialized skill development training. Not surprisingly, survey results varied by industry. The manufacturing industry tended to offer more training in group decision-making, individual planning and organization, and quality and statistical analysis. Both the insurance and finance industries, on the other hand, placed more value on interpersonal and communications skills, understanding the business, and career or personal development planning.

The apparent lack of demand for soft skills training in a technology-delivered format is matched by a correspondingly low supply of soft skills titles in multimedia format. Brandon Hall, a well-known technology-based training industry analyst, estimates that no more than 30% of total multimedia-based course titles are on soft skills training topics [57]. Earlier attempts to deliver technology-based soft skills training without the presence of a live instructor were disappointing, as they failed to engage their intended audiences and properly convey the complexity of the course material. However, there is a renewed interest in using multimedia-based materials, especially to supplement instructor-led programs. Technological improvements such as increased network bandwidth and computer processing speed, as well as improvements in instructional design and more realistic marketing of courses to properly set customer expectations, have made Web-based training feasible.

1.2.2.5 Providers of course hosting and customer care services

Data warehousing services include the use of computer servers to store course content and the provision of the delivery software required by users to access courses. Many companies, including Bell Canada, provide clients with access to "server farms" that are maintained by the service provider. Customer care services, also known as commerce service providers (CSPs), are provided by companies that run call center help desks for users of training products and provide e-commerce services such as billing and other customer transactions. IBM, Bell Canada, and others, particularly telecommunications companies, have had long experience in providing call center and electronic solutions to customers.

1.2.2.6 Providers of professional consulting services

Professional consulting services are provided by companies that specialize in training needs assessment, instructional design, train-the-trainer programs, and training evaluation services. Anderson is one of the largest providers of consulting services related to distance learning, but a large number of other consulting firms such as the Gartner Group, IBM, and independent consultants also provide this type of service.

1.3 Marketing Web-based training programs

To successfully launch a Web-based training program, courses must be properly designed, priced, distributed, and promoted. In classical marketing terms, this analysis is referred to as the “4 P’s”: product, place, price, and promotion. Sections 1.3.1–1.3.4 provide a brief overview of product, place, price, and promotion as they affect companies implementing Web-based training.

1.3.1 Product

The term “value proposition” refers to how a company’s product is differentiated from that of another company—in other words, the special characteristics that meet customer needs in a manner superior to other products in the product category. Product positioning, corporate image, and pricing are related to understanding the needs and mind-set of the consumer. Product positioning takes into account both the value proposition of the product and the company’s corporate image. Value proposition refers to the features that differentiate the product from those of competitors, such as flexibility, speed, and price. Corporate image is similarly tied to concepts that differentiate one company from another, as seen, for example, in descriptors such as premium provider, value provider, and low-cost provider.

Training departments must apply these concepts to properly align courses with customer needs and to promote the courses they offer. Criteria that can be used to rate the value of one Web-based training product over another include the following.

- Return on investment;
- Employee satisfaction;
- Competition in the market (i.e., whether consumers have many or few available substitutes for the product).

A competitive market is clearly emerging for both accredited and IT-related Web-based courses. In these areas, price as well as quality will be factors that differentiate one Web-based course from another. In other areas, where competition is not as strong, price will be a consideration only to the extent that potential consumers believe that the training is worth more than an alternative method of acquiring the desired information. For example, purchasing a book is a reasonable alternative to taking a course, especially when comparing a book with a self-paced course that offers no opportunity for student interaction with other students and an instructor. Table 1.1 summarizes the various categories of Web-based courses and the consumer demand for each type of course.

Web-based training has competition in classroom-based courses, classes offered through communications technologies such as videoconferencing and audiographic conferencing, and in computer-based training using CD-ROMs and other media. However, Web-based courses offer benefits such as low delivery costs; the ability to deliver interactive, multimedia-rich content; and the convenience to students

Table 1.1

Consumer Need for Web-Based Courses

Types of Web-Based Courses	Need (Based on Potential Return on Investment)
Accredited/certified courses	High
Special purpose courses (e.g., SAP)	High
IT-related courses	Medium
Product training for customers and resellers	Medium
General and soft skills courses	Low

of being able to take courses that are independent of time and location. Some of the key product features that competitors use to differentiate their Web-based courses are listed as follows.

- Price (and level of cost-effectiveness);
- Subject area;
- Quality of course content and effectiveness of instructional design;
- Amount of multimedia content;
- Special features such as integrated online training management;
- Level of interactivity for students and instructors;
- Ease of use;
- System compatibility and scalability of implementation, which allows a company to try a small number of Web-based courses and add more courses as required.

Many learning systems provide communications tools, including Centra Software's Symposium product, which allows live student groups to "break out" of a live training session to discuss particular problems and create common documents. Lotus Development and Microsoft also intend to release real-time collaboration features in software products directed at the distance-learning market. Microsoft's new release of Exchange will have collaboration tools, and Lotus Learning Space 3.0 is a real-time collaborative tool that also supports asynchronous communication [58].

Online training management is in demand by trainers and is a feature that system vendors use to distinguish their products from those of the competition. Some of the main companies providing software solutions for creating, delivering, and managing online courses include: Asymetrix Learning Systems (www.asymetrix.com); WBT Systems Inc. (www.wbtsystems.com); Macromedia (www.macromedia.com); Docent (www.docent.com); Allen Communications (www.allencomm.com); Oracle (www.oracle.com); Lotus Development Corp. (www.lotus.com); Centra Software (www.centra.com);

Interactive Learning International Corp. (www.ilinc.com); KnowledgeSoft (www.knowledgesoft.com); and WebCT Systems (www.webct.com).

1.3.2 Place

Although the nature of the Web lessens the importance of geographic location for service delivery to both domestic and international markets, the use of direct sales makes company location near the United States, where the greatest market for Web-based training is located, a logical choice. The choices for sales distribution channels include using a direct sales force, establishing retail outlets, and enlisting agents or resellers. Some of the considerations for choosing a sales channel include anticipation of customer satisfaction with the sales channel; consistency with product positioning, least cost for the level of service, least risk for the desired level of service, flexibility and control (e.g., over order cycle time, consistency, and accuracy), and feasibility. Most vendors of learning products rely primarily on their own sales force rather than resellers. There are several reasons that a direct sales force is an effective choice for learning products:

- The products are sufficiently new and complex to require a degree of consumer education before purchase.
- The cost benefit of Web-based training is a selling feature, and customers would typically need explanation and assistance in carrying out this type of analysis.
- Referring prospective purchasers to current users of a product for testimonials on how the courses were implemented and the effectiveness of the courses is an important part of the sales cycle.
- Most customers are located in large urban centers, making the use of a direct sales force feasible.

1.3.3 Price

The pricing of goods and services is based on market demand, competition, and cost of production. Product price needs to be aligned with product positioning and corporate image to maximize the effectiveness of the marketing effort and consumer demand. Monroe [59] identifies six factors that influence the willingness of customers to pay comparatively higher or lower prices for goods and services:

1. The perceived value of the product;
2. The intrinsic benefits of the product;
3. Consumer knowledge of the prices of competitive products;
4. The relative prices of similar products;
5. Customer expectation that current prices will not significantly increase;
6. The reputation of the product vendor.

As with classroom-based courses, consumers of Web-based courses are most likely to be willing to pay higher prices for courses that are both essential to the business and not widely available [60]. While companies would naturally prefer to recover their expenditures within a short period of time, competing companies with products that are easily substituted from the consumers' point of view ensures that pricing is competitive in the industry. Courses that respond to specific high-need areas for training and that are not easily accessible from alternative sources will command higher prices in the marketplace.

The pricing of Web-based courses varies greatly among vendors. Ganzel [61] has found four different pricing models used by vendors:

1. Charge per course hour;
2. Charge by monthly subscription;
3. Charge per fixed period (e.g., 10 weeks) for self-study courses;
4. Charge per element of course content (e.g., a video clip that can be included in a database).

Some course providers such as universities typically charge either the same price, or more, for Web-based courses as they do for similar classroom courses. Others such as Digital Think charge the same price as an equivalent classroom-based course, while companies such as Digital Education Systems charge one-quarter to one-fifth of the price of an equivalent classroom course. Naturally, these differing pricing schemes have led to highly varied prices. Courses available through America Online range in price from $19.95 for a six-month subscription to $200 for specialized courses such as those leading to Microsoft certification. Digital Think's prices range from $125 for a standard Microsoft Word course to $425 for advanced courses on emerging technology topics [62]. Other vendors are selling personal interest courses for $4.95 an hour, although $30–40 is a more common price. However, some technical certification courses can cost as much as $500 [63]. Chapter 5 evaluates in detail the pricing issues that apply to courses offered over the Web.

1.3.4 Promotion

As with any product, there is more demand for Web-based courses that are effectively promoted to potential end users. Promotion of courses can be done through a Web site as well as through more traditional formats such as advertising in periodicals and direct-mail campaigns and through customer contact with sales staff. Online Learning News [64] suggests that the best ways to promote online courses to potential in-house students are table tents, banners, and demonstrations. Promotion is recommended well before courses are available for registration. Weekly follow-up is also recommended after employees have enrolled in courses to encourage them to enroll in other online courses. Some suggested incentives are attractive "tickets" to subsequent courses and communication via mail, e-mail, and phone, with copies to employees' supervisors. Communications can outline the benefits of technology-based training and should be particularly aimed at employees who are leaders among their peers.

Carliner [65] recommends that in-house training departments identify an entire process for selling senior management on the benefits of online training. Suggestions for a marketing campaign include the following.

- A discussion about management's training concerns (for example, the length of training, the need for travel to classroom courses, the need for delivery of a consistent message, and an explanation of how online training would address those issues);
- A review of business issues including the expected financial bottom line results;
- Short communications such as e-mail messages that address the above issues in an interesting and concise manner;
- Hands-on demonstrations of technology-based learning products, following the communications program, that will help senior managers visualize how the technologies could be applied in their own organization;
- Presentation of a business case that clearly shows both the costs and the benefits of implementing online training;
- Recommendations for training products appropriate to the organization, whether off-the-shelf products such as Ziff-Davis Inc.'s LearnItOnline (www.learnitonline.com), DigitalThink Inc.'s online training (www.digitalthink.com), or the development of proprietary courses about company-specific information.

While some larger companies such as Bell Canada have even implemented training management systems, the majority of companies are still at the early adoption stage of Web-based training. While many companies are trying Web-based training, training programs are primarily focused on classroom courses. Although Web-based training is still experimental, its use is increasing. Table 1.2 summarizes the various marketing approaches at different stages of the market cycle [66].

1.4 Business process reengineering for Web-based training

BPR is the analysis, implementation, and evaluation of the various functions that comprise a business activity. After an organization

Table 1.2
Marketing Approaches to Web-Based Training

Market	Positioning	Promotion	Place	Price
Early adoption	Value is technology expertise applied to traditional training problems	Create demand for Web-based training through demonstration of cost benefit, convenience, improved service delivery, flexibility, and ease of use	Direct sales force	High
Mainstream	Value is in integrated solution (i.e., delivery of Web-based courses integrated with training management system, e-commerce features for course registration and payment)	Intimate knowledge of customer's business requirements Build customer references Manage established customer relationships	Direct sales force Resellers	Medium
Maturity	Fulfill on well-defined customer expectations	Wide acceptance, commoditization Time, place, price Value is operational excellence and low-cost delivery	Multivendor distribution	Low

Adapted from [66].

decides to implement Web-based training, it must reengineer its training function to accommodate changes. Some of the general principles for reengineering classroom training processes to include Web-based learning are the following.

- Defining the need for Web-based training;
- Establishing specific customer requirements and obtaining employees' and senior management's support for the proposed changes;
- Integrating appropriate technologies into the new business process;

- Setting clear goals for performance improvement that the reengineering process will accomplish;
- Managing change by carrying out a pilot project to test the feasibility of the solution;
- Evaluating the new process based on clear performance measurements.

1.4.1 Deciding between customization and off-the-shelf courseware

One of the important considerations from a BPR point of view is the decision whether to outsource course production. A survey of American companies found that human resources-related functions, including training, were outsourced more frequently than any other business activity, (84% compared with an average of 30% for non-HR functions) [67]. There is also a relationship between outsourcing and the trend toward performance management, in that performance-related training tends to be kept in-house while generic training—for example, on computer skills—is outsourced. This use of outsourcing keeps the focus of the training department on the business results that training brings rather than on resource-consuming but low-return functions like administration. The trend to outsourcing also strengthens relationships with colleges, universities, and training institutions, yet another motivation for repositioning the training department as a corporate university.

In the case of many large companies with in-house training departments, existing staff could potentially create Web-based courses instead of using external vendors to supply courses on contract. Table 1.3 summarizes some of the arguments for and against the outsourcing of Web-based course production.

A 1988 survey by Computer Reseller News found that 56% of respondents ranked the customization of training to reflect specific job functions as their most important concern [68]. Industry analyst Brandon Hall believes that soft skills training is the most common type of custom-designed training [69]. Eric Parks [70] suggests that, for the most part, companies will need to write their own course materials

Table 1.3
In-House Versus Outsourced Web-Based Course Production

In-House Course Production	
Pro	**Con**
Can control unit costs by controlling human resources and materials used and do not have to pay margin to outsource partner on top of unit costs	Must pay staff even if they do not always have enough work to stay fully occupied
Can control production quality	Production capabilities limited by expertise of hired staff and financial resources to hire additional expertise
Can develop unique product not generally available in the marketplace	Full costs of course development must be absorbed

Outsourced Course Production	
Pro	**Con**
Can contract for resources as required	Products developed by outsource partners equally available to competitors
Can hire different companies to take advantage of differing specialties (e.g., expertise in virtual reality for courses requiring simulation)	Unit prices generally higher when using outsource partner because outsource partner must add margin to its unit costs
Can offer wide variety of products by combining various offerings in market without absorbing full costs of development for each title	Available courses will not be tailored to fit specific needs of consumer Quality of products will need to be evaluated, before purchase if possible

because online course vendors aim for the mass market, offering generic courses that lack creativity. The high prices that online course vendors charge for customization approaches the cost of creating courses in-house. Companies can create a high-quality course through using a combination of in-house subject matter specialists and contracted help in instructional design for the Web and course production. Parks points out that by asking several multimedia companies, including smaller ones, to bid on the work, companies are likely to obtain a low price.

1.5 Barriers to the adoption of Web-based training

Web-based courses that include a significant amount of multimedia are expensive to produce, and although the reduced costs of delivery more than offset production costs when enough students need to be trained, this initial investment can intimidate companies considering the use of Web-based training. Few vendors are willing to take the risk of producing courses without a guaranteed market (hence the relatively high availability of courses on subjects such as Microsoft Office software, where a large potential market is assured). As well, potential students of technology-related courses are more likely to have access to computer equipment and possibly less likely to have psychological barriers to using this equipment to take courses than the general public. Although many companies are exploring the possibilities of Web-based training, their usage is still at the early adoption stage. Whether Web-based training will cross into the mainstream will depend on whether companies decide that the savings and flexibility in course delivery and online training management advantages outweigh factors such as the following.

- High initial costs;
- The need to ensure that users have appropriate computer equipment and network capacity;
- The need to reengineer current training processes.

However, there may also be risks associated with not adopting Web-based training. Changing technologies and market globalization have placed companies under greater stress, and the need to deliver fast, effective, and easily updated training materials is more important than ever before. Along with an increasing need for training is the conflicting requirement to operate companies with fewer costs, in order to remain competitive. Training departments are typically viewed by companies as a cost center, and in difficult financial times, they are highly subject to cutbacks. Classroom training is expensive to deliver, and printed course materials are expensive and time-consuming to produce. Web-based training allows training departments to deliver quality training efficiently and cost-effectively, thereby minimizing the effects of cutbacks to training programs.

1.6 Conclusion

Training plays an important role in a company's ability to compete in knowledge-intensive global markets. However, simply increasing expenditures on traditional classroom training does nothing to ensure that the training is meeting the needs of employees and the organization. Web-based training creates many efficiencies, such as a reduction in the cost of training delivery, speed in updating course materials, flexibility in time and geographic location of course delivery, and the ability to manage training to ensure that corporate and individual needs are met. Many companies are using Web-based training, but most are still at an experimental stage with this technology. Whether popular opinion decides that the advantages outweigh the disadvantages of high initial costs, the complications of ensuring that equipment and networks are adequate and the need for reengineering training practices will determine whether Web-based training enters the mainstream.

References

[1] National Alliance of Business Inc., "Company Training and Education: Who Does It, Who Gets It and Does It Pay Off?" *Workforce Economics*, Vol. 3, No. 2, June 1997.

[2] Benson, George, "Is Training Different Across the Border?" *Training and Development*, Vol. 51, No. 10, October 1997, pp. 57–58.

[3] National Alliance of Business Inc., "Company Training and Education: Who Does It, Who Gets It and Does It Pay Off?" *Workforce Economics*, Vol. 3, No. 2, June 1997; *Profile of Post-Secondary Education in Canada* [Ottawa]: Education Support Branch, Human Resources Development Canada, 1994.

[4] Edupage, "Tech Workers Good As Gold," June 2, 1999, (http://educause.unc.edu/).

[5] Masie, Elliott, "Taking the Business Pulse of Online Learning and Training," *Computer Reseller News*, No. 833, March 15, 1999, p. 52.

[6] Imel, Susan, *Web-Based Training: Trends and Issues Alerts,* Washington, D.C.: Office of Educational Research and Improvement, 1997.

[7] Lakewood Publications Inc., "Training Magazine's Industry Report, 1998," *Training,* Vol. 35, No. 10, October 1998, pp. 43–45.

[8] Benson, George, "Is Training Different Across the Border?" *Training and Development,* Vol. 51, No. 10, October 1997, pp. 57–58.

[9] Tucker, Brian, *Research Into the Use of Intranets for Training*, Wirral, England: Forum for Technology in Training, 1997.

[10] Benson, George, "Is Training Different Across the Border?" *Training and Development,* Vol. 51, No. 10, October 1997, pp. 57–58.

[11] Lakewood Publications Inc., "Still Waiting for Training Software," *Online Learning News,* Vol. 1, No. 22, August 31, 1998.

[12] Lakewood Publications Inc., "'Wow' Moments at the Show," *Online Learning News,* Vol. 1, No. 26, September 29, 1998.

[13] Benson, George, "Is Training Different Across the Border?" *Training and Development,* Vol. 51, No. 10, October 1997, pp. 57–58.

[14] Edupage, "Commonplaces' Web Strategy Targets Students," May 10, 1999, (http://educause.unc.edu/).

[15] Cole-Gomolski, Barb, "Sears Slices IT Turnover to 9%," *Computerworld,* Vol. 32, No. 47, November 23, 1998, pp. 39–40.

[16] Torode, Christine, "Microsoft Debuts Seminar Online," *Computer Reseller News,* No. 792, June 8, 1998, pp. 49–50.

[17] Alexander, Steve, "Web Certifications Abound," *InfoWorld,* Vol. 20, No. 15, April 13, 1998, pp. 91–94.

[18] Zieger, Anne, "Vendors Push Web Certification," *InternetWeek,* No. 730, August 31, 1998, pp. 28–29.

[19] Cornell University, "Cornell Medical College Offers Internet Delivery of Grand Rounds," *Health Care Strategic Management,* Vol. 16, No. 11, November 1998, pp. 4–5.

[20] Apicella, Mario, "Technology for the Classroom and the Battlefield," *InfoWorld,* Vol. 20, No. 36, September 7, 1998, p. 77.

[21] Smith, Stephen E., "Ongoing Medical Education on the Web," *Information Today,* Vol. 15, No. 10, November 1998, pp. 24–25.

[22] Lakewood Publications Inc., "Certification Online Speeds Needed Learning for Nurses and Financial Pros," *The Lakewood Report on Technology for Learning,* Vol. 5, No. 6., June 1999, p. 4.

[23] McCausland, Richard, "Online CPE Explodes," *Accounting Technology*, Vol. 15, No. 2, March 1999, pp. 24–28.

[24] Ludorf, Carol A., "www.truklink.com: A Door To Learning," *Fleet Equipment*, Vol. 25, No. 2, February 1999, p. 20.

[25] Wilder, Clinton, "Training on the Internet," *InformationWeek*, No. 679, April 27, 1998, p. 130.

[26] Torode, Christine, "Consulting, Custom Solutions Add Zest to Vanilla IT Training," *Computer Reseller News*, No. 826, January 25, 1999, pp. 45–48.

[27] Edupage, "Is Education the Next Online Money-Maker?" June 9, 1999, (http://educause.unc.edu/).

[28] Fisher, Anne, "Getting a BA Online, and Finding a Female Mentor," *Fortune*, Vol. 139, No. 2, February 1, 1999, p. 144.

[29] Edupage, "Passing Without Distinction," June 19, 1999, (http://educause.unc.edu/).

[30] National Alliance of Business Inc., "Company Training and Education: Who Does It, Who Gets It and Does It Pay Off?" *Workforce Economics*, Vol. 3, No. 2, June 1997.

[31] University of North Carolina at Chapel Hill, Center for Instructional Technology, "Study on Distance Learning Programs in Higher Education," *CIT INFOBITS*, No. 12, June 1999 (http://www.unc.edu/cit/infobits/infobits.html).

[32] Kastelein, Barbara, "Professional Development," *Business Mexico*, Vol. 9, No. 3, March 1999, pp. 46–49.

[33] Lakewood Publications Inc., "H.R., I.T., K.M. and E.P.S.S.—What's That Spell?" *Online Learning News*, Vol. 2, No. 6, May 11, 1999.

[34] Stuart, Anne, "Beyond the Campus," *CIO*, Vol. 11, No. 22, September 1, 1998, pp. 30–40.

[35] Edupage, "Passing Without Distinction," June 19, 1999 (http://educause.unc.edu/).

[36] Edupage, "Western Governors U. Seeks an $8-Million Federal Appropriation," May 19, 1999 (http://educause.unc.edu/).

[37] Edupage, "Passing Without Distinction," June 19, 1999 (http://educause.unc.edu/).

[38] Edupage, "Keeping Company With the Campus," April 26, 1999 (http://educause.unc.edu/).

[39] Edupage, "It's Academic," June 2, 1999 (http://educause.unc.edu/).

[40] Wilson, Linda, "Virtual Training Saves $800K in First Year," *Computerworld*, Vol. 32, No. 12, March 23, 1998, p. S11.

[41] Benson, George, "Is Training Different Across the Border?" *Training and Development*, Vol. 51, No. 10, October 1997, pp. 57–58.

[42] Edupage, "Keeping Company With the Campus," April 26, 1999 (http://educause.unc.edu/).

[43] Machlis, Sharon, "Schwab Wants Users to Stay Awhile," *Computerworld*, Vol. 32, No. 47, November 23, 1998, pp. 45–48.

[44] O Sullivan, Orla, "'Webizing' Learning," *US Banker*, Vol. 108, No. 9, September 1998, p. 20.

[45] Reynolds, Cynthia, "In the Key of Green," *Canadian Business*, Vol. 71, No. 19, November 27, 1998, p. 166.

[46] Edupage, "Distance Learning Is No Substitute for Real-World Education," May 21, 1999 (http://educause.unc.edu/).

[47] Edupage, "A Shortage of Tech Workers?" May 10, 1999 (http://educause.unc.edu/).

[48] Edupage, "Campus Pipeline Leads Advertisers to a Demographics Goldmine: Students," May 24, 1999 (http://educause.unc.edu/).

[49] Edupage, "Making New Markets," May 21, 1999 (http://educause.unc.edu/).

[50] MacNeil, Donald R., *Education Services Market*, Ottawa: Bell Canada, 1996.

[51] Lakewood Publications Inc., "Training Magazine's Industry Report, 1998," *Training*, Vol. 35, No. 10, October 1998, pp. 43–45.

[52] Kalin, Sari, "Internet Class Act," *CIO*, Vol. 11, No. 18, July 1, 1998, pp. 30–34.

[53] Torode, Christine, "More Bang for the Buck in Internet Training," *Computer Reseller News*, No. 833, March 15, 1999, pp. 191–196.

[54] Bassi, Laurie J., Scott Cheney, and Mark Van Buren, "Training Industry Trends 1997," *Training and Development*, Vol. 51, No. 11, November 1997, pp. 46–59.

[55] Roberts, John, "K–12 Classrooms Are VAR Gold Mine," *Computer Reseller News*, No. 814, November 2, 1998, pp. 96–98.

[56] Salopek, Jennifer, "Arrested Development," *Training and Development*, Vol. 52, No. 9, September 1998, p. 65.

[57] Barron, Tom, "The Hard Facts About Soft-Skills Software," *Training and Development*, Vol. 52, No. 6, June 1998, pp. 48–51.

[58] Edupage, "Lotus, Microsoft Vie for Distance Learners," May 28, 1999 (http://educause.unc.edu/).

[59] Monroe, Kent B., *Pricing: Making Profitable Decisions*, New York: McGraw-Hill, 1990.

[60] Carliner, Saul, "What To Pay, What To Charge," *Online Learning News*, Vol. 1, No. 47, February 23, 1999.

[61] Ganzel, Rebecca, "What Price Online Learning?" *Training*, Vol. 36, No. 2, February 1999, pp. 50–54.

[62] Kalin, Sari, "Internet Class Act," *CIO*, Vol. 11, No. 18, July 1, 1998, pp. 30–34.

[63] Carliner, Saul, "What To Pay, What To Charge," *Online Learning News*, Vol. 1, No. 47, February 23, 1999.

[64] Lakewood Publications Inc., "Luring Learners Online," *Online Learning News*, Vol. 2, No. 3, April 20, 1999.

[65] Carliner, Saul, "Pitching Online Learning to Management," *Online Learning News*, Vol. 1, No. 11, June 15, 1998.

[66] Moore, Geoffrey A., *Crossing the Chasm: Marketing and Selling High-Tech Products to Mainstream Customers*, New York: HarperBusiness, 1995.

[67] Bassi, Laurie J., Scott Cheney, and Mark Van Buren, "Training Industry Trends 1997," *Training and Development*, Vol. 51, No. 11, November 1997, pp. 46–59.

[68] Torode, Christine, “VARs Step Up Web-Based Training,” *Computer Reseller News*, No. 826, January 25, 1999, pp. 1–6.

[69] Barron, Tom, “The Hard Facts About Soft-Skills Software,” *Training and Development*, Vol. 52, No. 6, June 1998, pp. 48–51.

[70] Lakewood Publications Inc., “What Teller Training Teaches All Online Trainers,” *Online Learning News*, Vol. 1, No. 23, September 8, 1998.

2

Cost-Benefit Analysis of Web-Based Training: Case Study of the Bell Online Institute[1]

2.1 Introduction

Web-based training is receiving a great deal of interest in academia and private industry, and cost analysis has become increasingly important. Many universities such as the University of Phoenix, the University of Maryland, and Athabasca University, to name a few, have multi-

1. Reprinted with the permission of Pennsylvania State University, *American Journal of Distance Education,* Vol. 13, No. 1, pp. 24–44.

million-dollar budgets for Web-based teaching of part-time students. In private industry, Bell Canada is a good example of how companies are using Web-based learning for both in-house employee training and as a service product offered to customers who want to apply distance learning in their own organizations. In both cases, the cost benefit of Web-based training must be established.

Internally, Bell Canada is interested in using Web-based training because it is a cost-effective and convenient alternative to classroom training. Since Web-based training is also a customer product, it is necessary to demonstrate its value in order to interest customers in using this type of training model. For these reasons, Bell Canada undertook a pilot project to examine the cost-effectiveness of Web-based training.

The literature in the area of cost-benefit analysis of Web-based distance learning is sparse. The costing methodology discussed in this chapter was developed by the authors and tested in the Bell Canada case study. The results of the cost-benefit analysis provide further information about the economic benefits of Web-based training.

2.2 Measurements of financial performance

Although few cost-benefit analyses of Web-based training are available, standard measures of financial performance apply to this type of study. Two common measures are the break-even point—the point at which costs are recovered—and return on investment (ROI), which illustrates the economic gain or loss from having undertaken a project.

2.2.1 Break-even number of students

To offset the high fixed costs of Web-based courses, a certain number of students must be trained at a delivery cost per student of less than that of the delivery cost per student for classroom training. The number of students that offsets the fixed costs of Web-based training is the break-even point. In Figure 2.1, the total fixed costs for classroom and Web-based courses are represented by the intercepts of the vertical axis. The cost of course delivery per student is represented by the slope of each line. Since Web-based courses cost more to develop, the line

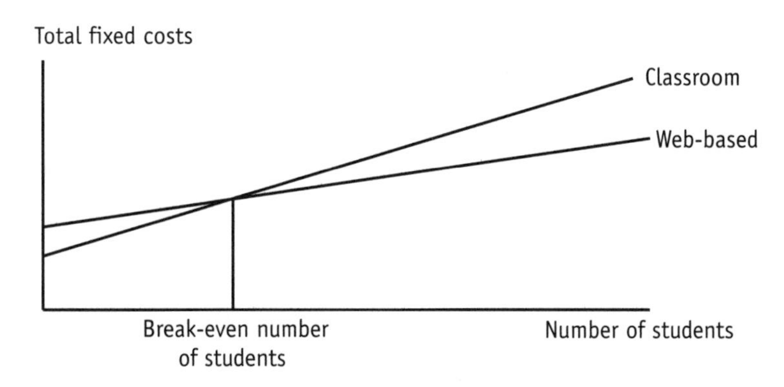

Figure 2.1 Break-even number of students.

meets the vertical axis at a higher point than that for classroom costs. However, since Web-based courses cost less to deliver per student, the slope of the line is more gradual. The point at which the lines cross is the break-even number of students needed to recover the higher fixed costs of Web-based delivery.

2.3 Cost comparison: a background

In comparing Web-based distance education and classroom course delivery, there must be a clear basis for measuring output. It can be argued that the true measure of training output is the information and skills acquired by students. However, many studies report that there is no significant difference in the amount of learning that students acquire through classroom training and distance learning [1]. For the present costing analysis, the basis for comparison is the cost of development and delivery of the courses themselves. As with any mode of course delivery, the amount, type, and format of the learning content as well as its presentation to students must be appropriate to the subject and the intended audience. Differences in course development and delivery costs related to variations in instructional design were not examined here, but this would make an interesting topic for future study. This analysis assumes that the learning outcomes for the Web-based courses in this study would be the same if the courses had been delivered in a classroom, and the basis for comparison is the cost of delivering an equivalent course in the classroom.

Costs are divided into fixed and variable costs in order to compare technology-enabled learning with traditional classroom delivery. Fixed costs are defined as costs that remain the same regardless of the output. Variable costs are those that vary directly with the amount of output. Thus, variable costs increase with the number of students while fixed costs are incurred before a course can be offered to the first student. Clearly, costs that would have been incurred even if a course were delivered in a classroom—so-called sunk costs—should be ignored in costing Web-based training. For example, if a course were previously offered in the classroom, and no further research and development for course materials needed to be done, the cost of creating the intellectual property for the course would be a sunk cost. Likewise, the cost of feasibility studies should not be included as a costing element.

In their analysis of distance education using videoconferencing-based delivery at the University of Ottawa, Aubé, Chilibeck, and Wright [2] established the fixed costs for videoconferencing as: videoconferencing equipment, technicians' salaries for running the equipment, installation costs, and the fees for basic telephone lines. Variable costs for videoconferencing delivery of distance courses were understood to be fees for long distance network usage, shipping charges for supplementary print materials, honoraria for professors, and salaries paid for the preparation of course materials. The break-even point—found to be 22 courses with the videoconferencing configuration used at the University of Ottawa—was the financial measure used in the analysis.

Trevor-Deutsch and Baker [3] reported on the costing of a course with multimedia elements delivered via videoconferencing in an academic setting. They considered the following costs involved with traditional classroom delivery: the instructor's salary and benefits, the number of courses the instructor teaches (i.e., when instructors are paid on a per-course basis), the costs of course development, course materials, administrative support, classroom overhead, and any additional time the instructor spent on the course, such as for purposes of grading and meeting with students. Costs for the same course delivered at a distance included equipment costs and course development costs. The break-even point was calculated at 331 students. Future courses, using existing equipment and some of the development features of the pilot course, were estimated to have a break-even point of 82 students, a 75% reduction in the break-even point of the pilot.

Another study by Brandon Hall compared CD-ROM-based learning to classroom learning in a high-tech industry setting [4]. Costs for traditional classroom delivery included development time, classroom overhead costs, instructors' costs, and travel costs for the participants. The CD-ROM training costs included equipment and course development time. Development time was high for the CD-ROM course, totaling $1,205,394 over three years. However, the classroom-delivery costs were also high, primarily due to high travel costs for the students and the length of time spent away from the job. Costs over the three-year pilot period were 47% less than the classroom-based course. The payback period was found to be 15 months, and the internal rate of return—the discount rate that makes the net present value of the investment equal to zero—was 61%. The break-even number of students needed to recover development costs was not calculated. The course material was compressed by 60%, from four days in the classroom to 11.2 hours for the CD-ROM version. The reduction in training time was attributed to improved instructional design, the ability of students to test their knowledge and bypass some sections of the course, and the variety of instructional models available to students, such as text and animation, which contributed to more effective learning.

Several other publications are relevant to the discussion of Web-based costing analysis. In a 1998 article, Hall itemized the current contractor rates for the types of experts required for Web-based course production. In his 1995 book, Bates [5] also discusses the costing of distance education courses, although the Web is not included in his analysis. A more recent study by Bartolic-Zlomislic and Bates examines the costing of Web-based courses in a university setting [6]. The methodology used in that study included an assessment of capital and recurrent costs, production and delivery costs, and fixed and variable costs, as well as performance-driven, value-driven, and societal benefits.

Another report by the Higher Education Information Resources Alliance [7] primarily examines noneconomic benefits of the use of information technology in an academic setting.

Several bibliographies on both ROI and evaluation have been compiled by the American Society for Training and Development [8] and are available on the society's Web site at http://www.astd.org/. An interesting three-part series in *Training and Development* by Phillips [9] discusses various issues surrounding the measurement of training effectiveness. While all of these studies are useful for a comparative look at methods of determining the costs and benefits of distance education, few reports specifically deal with the economic analysis of Web-based courses. This study takes a new approach in looking at cost-benefit analysis for the new delivery method of the Web.

2.4 Costing methodology

Web-based training has become a widely explored topic in the education and training literature, but there are still few comprehensive and tested costing methodologies available for use by educators, trainers, and businesspeople who need to make decisions about the cost-effectiveness of Web-based training for their own organizations. This study hypothesizes that there are several key design elements that must be costed in a majority of Web-based training projects. These costs are divided into fixed capital costs and variable operating costs.

Capital costs include the server platform shared over all the courses mounted on that server, as well as the cost of the content development, which is shared over all the students taking that course. Content development includes six items:

1. Instructional and multimedia design;

2. The production of text, audio, video, graphics, and photographs;

3. The development of authoring and delivery software, or the cost of licensing commercial software;

4. The technical integration, modification, and testing of course content;
5. Student and instructor training;
6. Course testing.

Operating costs represent the costs for time that students and instructors spend using the courses.

These costs are analyzed to determine the costs per course, the costs per phase of development, the costs per student, and the costs per mode of delivery (i.e., synchronous and asynchronous). The discussion of ratio analysis later in this chapter provides an evaluation of the break-even number of students required to recover course development costs and ROI over five years. The costs of producing future courses using the same instructional design or mode of delivery can be estimated from the results of this study.

To verify the costing methodology, it is necessary to determine whether some of these costing elements are more important than others. In other words, do some of the elements dominate others, or are some elements insignificant? To test the hypothesis that all these elements are critical to the costing analysis of Web-based course design and delivery, the costing methodology is applied to the following case study.

2.5 Bell Online Institute case study

The Bell Online Institute (BOLI) is the Bell Canada business unit that delivers Web-based training to Bell Canada employees. BOLI also functions as a testing ground for various Internet-based learning platforms used for delivery of training at Bell Canada. A separate business unit, the Bell Institute for Professional Development (BIPD), provides traditional classroom course delivery and oversees all internal training.

The BOLI case study was undertaken to measure the cost and evaluate the effectiveness of training delivered on four different Web-based learning platforms:

- WebCT (http://www.webct.com/), developed at the University of British Columbia for asynchronous training;
- Mentys (http://www.globalknowledge.com/) from Global Knowledge Network, for asynchronous training;
- Pebblesoft (http://www.pebblesoft.com/), for asynchronous training;
- Centra Symposium (http://www.centra.com/), for synchronous training.

Three courses designed for delivery on these learning platforms were produced by three independent vendors. All of the courses were on telecommunications topics: transmission control protocol/Internet protocol (TCP/IP) (using the WebCT and Mentys platforms), frame relay (using the Pebblesoft platform), and routing (using the Centra Symposium platform). The courses were estimated to be equivalent to two-day classroom courses. The pilot courses were delivered to engineers working in Bell Canada's Advanced Communication Systems group.

Three of the learning platforms, WebCT, Mentys, and Pebblesoft, present course materials asynchronously. That is, course materials reside on a server on an ongoing basis and may be accessed at the student's convenience. The frame relay course on the Pebblesoft learning platform was authored in French. The fourth platform, Symposium, is a synchronous learning platform using 28.8-Kbps delivery over the Internet or an intranet. The system supports text, graphics, and animation to present course materials. Shared features for system users of Symposium include audio communication among the course participants and the instructor, an electronic whiteboard, an Internet browser, and a live text chat room.

2.6 Cost analysis: Fixed costs

This portion of the analysis helps determine whether the high fixed costs associated with providing learning in a technology-enabled format are justified when compared with the costs of traditional classroom delivery already provided through BIPD. The actual billed rates reported by the vendors are used in the cost charts found throughout the chapter. All costs are reported in Canadian dollars.

2.6.1 Capital costs

2.6.1.1 License fees for learning platform software

The price of a learning platform depends on the number of people who will be using the software and ranges widely from vendor to vendor. An upgrade cost of 10% per year is assumed. Using 10% of the costs as an annuity, the present value of the upgrades and the original purchase price are shown in column three of Table 2.1. The cost of the platform must be amortized over the total number of courses that will use the platform. BIPD presently has 150 independent-study courses on CD-ROM or diskettes. It is assumed that all of these courses could potentially be taught using any of the asynchronous platforms. There are presently 700 classroom-based courses offered through BIPD. It is estimated that 10% of these could potentially be offered using the Centra Symposium synchronous platform. Table 2.1 summarizes the costs associated with learning platforms.

Table 2.1
Learning Platform Costs

Platform	License	Upgrade Cost	Cost Per Course
WebCT (asynchronous)	$3,000	$4,103 ($3,000 + $1,103)	$27 per course ($4,103/150)
Mentys (asynchronous)	$150,000	$205,161 ($150,000 + $55,161)	$1,368 per course ($205,161/150)
Pebblesoft (asynchronous)	$175,000	$239,348 ($175,000 + $64,348)	$1,596 per course ($239,348/150)
Symposium (synchronous)	$35,000	$47,871 ($35,000 + $12,871)	$684 per course ($47,871/70)

2.6.1.2 Hardware

In this study, it is assumed that a server must be purchased to offer the course but that client computers are already available on the desktops of the employees participating in the course. The purchase price of the server was estimated to be $15,000. The total number of courses, as established previously, is 220 (150 asynchronous + 70 synchronous). It is assumed that not all courses would need to run on the server at one time. By amortizing the cost of the server over the number of courses, the cost of the server per course equals $68 ($15,000/220).

During the pilot study, staff time was required to set up desktop machines—for example, the installation of browser plug-ins—to take advantage of the multimedia aspects of the courses. However, this cost was not included in the study since newer versions of the learning platforms would automatically install the software needed for the desktop computers.

2.6.1.3 Total capital costs per course

The capital cost per course is equal to a server cost of $68 ($15,000/220) plus the platform costs per course for each different platform:

- Synchronous routing course: $752 ($68 + $684 for Symposium);
- Asynchronous TCP/IP course using WebCT: $95 ($68 + $27);
- Asynchronous TCP/IP course using Mentys: $1,436 ($68 + $1,368);
- Asynchronous frame relay course using Pebblesoft: $1,664 ($68 + $1,596).

2.6.2 Phase 1: content development

Content development variables include: instructional design; multimedia design; the production of text, audio, motion video, graphics, and photos in machine-readable format; course authoring; software development; integration of content and testing; modification/

adjustment; training; and course testing. Only one of the courses, TCP/IP, included motion video elements.

2.6.2.1 Developer salaries

The billed hourly rates for the independent contractors were used in the analyses found throughout this chapter. These are the rates that customers of these companies are willing to pay for Web-based course development.

2.6.2.2 Differences in synchronous and asynchronous course development

Significant differences were found between the asynchronous courses and the synchronous course regarding course development costs. The asynchronous courses were similar in the types of tasks performed by developers and the total amount of time spent on course development. The synchronous course, however, required far less development time, primarily due to less use of multimedia. For example, the synchronous course contained no audio, video, or photographic elements, and also had fewer graphics. An average of 1,321 development hours were spent on the asynchronous courses, while only 144 development hours (89% fewer) were required for the synchronous course. Since development costs depend on the number of hours required, the synchronous course cost much less to develop.

2.7 Cost analysis: Variable costs

2.7.1 Phase 2: Usability testing

Phase 2 of the BOLI pilot project involved course testing in which Bell Canada employees participated in training using the Web-based courses. This represented the total variable costs for the project. Since Bell Canada must pay employees for the time they spend in training, student salary costs are a significant factor in this costing analysis. More time spent in course delivery translates into higher student salary costs and less cost savings. The salary costs of students and, for the synchronous course, for both students and the instructor, were analyzed to compare the delivery costs per student. Figure 2.2 gives an overview of the delivery costs for all courses. These costs are described in more detail in the following section.

Table 2.2 gives a summary of the fixed and variable costs for each Web-based course in the study as well as the baseline classroom course

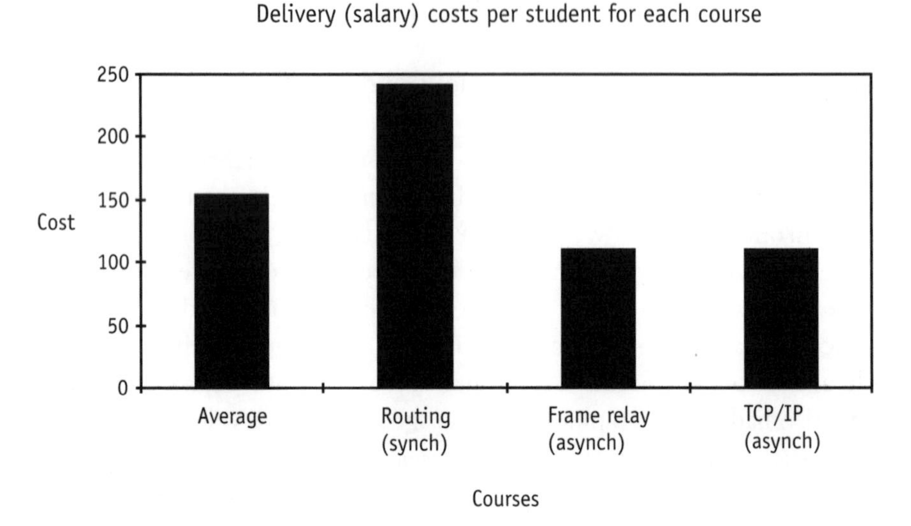

Figure 2.2 Delivery (salary) costs per student for each course.

Table 2.2

Fixed and Variable Costs for Web-Based Courses

Course	Total Fixed Costs	Total Variable Costs
Classroom	$28,600	$1,286 ($600 + 616 + 70)
Routing	$24,332 ($23,580 + 752)	$242 ($176 + 66)
Frame relay	$88,364 ($86,700 + 1,664)	$110
TCP/IP on WebCT	$158,708 ($158,613 + 95)	$110
TCP/IP on Mentys	$160,049 ($158,613 + 1,436)	$110

costs. A detailed analysis discussing how these numbers were derived is found in the following section.

2.8 Cost analysis of courses

2.8.1 Classroom courses

2.8.1.1 Life span and duration of courses

The life span of the courses is five years, meaning that no upgrade of course material is expected within a five-year period. The duration of classroom courses is 14 hours. Although the actual time spent in the classroom is approximately 12 hours, the course participants are paid for two normal seven-and-a-half hour workdays.

2.8.1.2 Costs of classroom training at the Bell Institute for Professional Development

Although none of the Web-based courses in this pilot project had been offered previously in the classroom, it was estimated that the courses were equivalent to two-day classroom courses. Therefore, the costs of delivering these courses in a classroom setting can be calculated by using the figures for delivering a typical two-day classroom course at BIPD.

2.8.1.3 Course development costs

For each day of classroom training BIPD requires 20 days of development (20:1). BIPD charges two different development rates depending upon whether or not BIPD will deliver the courses and have the opportunity to recover some development costs from future delivery. If BIPD delivers the course, there is a charge of $715 per day for course development to the business unit requesting the course. If BIPD does not deliver the course, there is a charge of $995 per day for course development. Therefore, the development of a two-day classroom course similar to the ones in the pilot project would cost the customer a total of $28,600 ($715 × 40 days).

2.8.1.4 Tuition fees

In-house tuition fees would also be charged to the participants' department to recover BIPD's normal operating costs. Classroom courses at BIPD have, on average, ten participants. The tuition charged to each participant is $250, and a further $50 per person is charged to cover such things as the cost of the instructor, course materials, and refreshments. This equals a cost per participant of $300 per day. Therefore, for a two-day course like those in the pilot, a cost of $600 per participant would be charged by BIPD to the business unit in which the participant works.

BOLI did not charge tuition to participants in the pilot. However, tuition charges for Web-based courses are anticipated with the full implementation of Web-based training to Bell employees. The costing information in this study will be taken into account to determine an appropriate price to charge business units in the company.

2.8.1.5 Salary costs for course participants

Twenty Bell Canada engineers were enrolled in the synchronous course, each earning an average annual salary of $57,700. The loaded annual salary including vacation, pensions and benefits, a portion of a supervisor's salary, office space, and office supplies and services is $85,500 per year, or $44 per hour.

Each of the three pilot courses was equivalent to a two-day classroom course. Each participant was paid for 14 hours of work for this period, at an average loaded hourly rate of $44. Therefore, the salary cost of offering classroom courses was $616 per participant.

2.8.1.6 Travel costs

Approximately 10% of employees taking courses at BIPD travel to a different city to take the course. At a typical travel cost of $350 per day, which includes airfare, one night in a hotel, and travel allowances, a two-day course costs Bell Canada $700 per student who travels. If 10% of the students travel, the travel cost per student is $70.

2.8.2 Web-based courses

2.8.2.1 Life span of courses

The life span of the Web-based courses is also five years. The duration of the Web-based courses is two-and-a-half hours for asynchronous courses and four hours for synchronous courses.

2.8.2.2 Course compression

Compression ratios for the Web-based courses are 79% for asynchronous and 67% for synchronous courses. Hall [10] points out that there is strong evidence that CD-ROM-based multimedia training requires less time than classroom training. Reported course compression ratios range from 20% to 80%, with 40% to 60% being the most common for CD-ROM training. In the BOLI pilot, course compression ratios were at the high end of the reported ranges in the Hall study for both synchronous and asynchronous courses. The Web-based asynchronous courses lasted two-and-a-half hours, while comparable classroom versions would have lasted 12 instruction hours. Course compression was, therefore, 79% for both asynchronous courses. The synchronous course lasted four hours, while a comparable course offered in a classroom would have lasted 12 instruction hours. This equaled a compression ratio of 67% for the synchronous course.

2.8.2.3 Salary costs for course participants

Synchronous Course For one engineer to take the four-hour synchronous course, salary costs totaled $176 ($44 × 4). There was also a live instructor present during the full four hours of the course. Theoretically, the Centra Symposium system can accommodate 50 or more participants at one time. However, this large number of concurrent participants would seriously limit the opportunities for interaction among the students and the instructor. For this study, it was assumed that the routing course would be taught in groups of 10. To calculate the cost of the instructor's salary per student, an hourly rate of $164 was used—the billed rate reported by the course vendor. This equaled a cost of $66 per student ($164 × 4 hours/10 students).

Asynchronous Courses The asynchronous courses each had 29 participants, who were also Bell Canada engineers. Based on the average loaded rate of pay, the cost for one engineer to complete a two-and-a-half hour asynchronous course was $110 ($44 × 2.5).

There are also opportunity costs in having employees spend time away from their work to take courses. For example, the completion of a project may be delayed due to time spent in training, and that delay may have unforeseen consequences for the organization's business. One advantage of asynchronous Web-based training as opposed to either synchronous or classroom training is that these problems are alleviated due to greater flexibility in scheduling students' time. However, although student salary costs are one measure of opportunity costs, total opportunity costs are almost impossible to determine and, therefore, have not been included here.

2.9 Web-based course development costs

2.9.1 Asynchronous courses

2.9.1.1 Frame relay course

The frame relay course is one of two courses delivered on an asynchronous platform. The course vendor had difficulty in using Pebblesoft because the initial software was delivered without French character support. This caused an estimated 24% increase in development time. If the project had been completed without the software problem, the hours spent on course development would have equaled 932 hours, and development costs at $562.50 per day would have totaled $69,900. Table 2.3 gives the total cost of each developer task at the billed rates reported by the course vendors and also summarizes the number of hours spent on each task, including the percentage of the total that this represents.

2.9.1.2 TCP/IP course

Another vendor produced an asynchronous course on the topic of TCP/IP. Two different platforms were used to deliver the course: Global Knowledge Network's Mentys system and WebCT, developed

Table 2.3

Frame Relay Course (Asynchronous): Cost and Time per Developer Task

Developer Task	Cost	Percentage of Time
Training	$1,200	1.4% (16 hrs.)
Tests	$9,975	11.5% (133 hrs.)
Instructional design	$18,525	21.4% (247 hrs.)
Multimedia design	$13,125	15.1% (175 hrs.)
Text production	$6,225	11.5% (133 hrs.)
Audio production	$1,875	2.2% (25 hrs.)
Graphics production	$12,300	14.2% (164 hrs.)
Photo production	$1,575	1.8% (21 hrs.)
Authoring/software development	$6,075	7% (81 hrs.)
Content integration	$7,800	9% (104 hrs.)
Testing integration	$2,550	2.9% (34 hrs.)
Modification/adjustment integration	$5,475	63% (73 hrs.)

Note: Total cost at $562.50 per day = $86,700. (All costs are reported in Canadian dollars.) Total of 1,156 hours.

at the University of British Columbia. A staff member of BIPD also participated in course development by designing the content architecture for the instructional design plan. The TCP/IP course included a five-minute video segment and was the only course that included video. Table 2.4 gives the total cost per developer task as well as the number of hours spent on each task and the percentage of the total time spent.

2.9.2 Synchronous course

2.9.2.1 Routing course

The routing course was delivered on the Centra Symposium platform. The pilot course was delivered to 20 Bell Canada engineers from Quebec City, Montreal, Ottawa, Toronto, and Hamilton. The course was offered in two parts, one in the morning and one in the afternoon. The morning session lasted two and a half hours and the afternoon session

Table 2.4

TCP/IP Course (Asynchronous): Cost and Time per Developer Task

Developer Task	Cost	Percentage of Time
Training	$7,893	5% (74 hrs.)
Video production	$26,667	16.8% (250 hrs.)
Instructional design	$36,800	23.2% (345 hrs.)
Multimedia design	$9,173	5.8% (86 hrs.)
Text production	$747	0.1% (7 hrs.)
Audio production	$4,053	2.6% (38 hrs.)
Graphics production	$31,253	19.7% (293 hrs.)
Authoring/software development	$12,907	8.1% (121 hrs.)
Content integration	$19,093	12% (179 hrs.)
Testing integration	$3,733	2.4% (35 hrs.)
Modification/adjustment integration	$6,293	4% (59 hrs.)

Note: Total cost at $800 per day = $158,612. (All costs are reported in Canadian dollars.) Total of 1,487 hours.

Table 2.5

Routing Course (Synchronous): Cost and Time per Developer Task

Developer Task	Cost	Percentage of Time
Training	$2,620	11.1% (16 hrs.)
Tests	$655	2.8% (4 hrs.)
Instructional design	$17,849	75.7% (109 hrs.)
Multimedia design	$491	2.1% (3 hrs.)
Text production	$1,310	5.6% (8 hrs.)
Graphics production	$655	2.8% (4 hrs.)

Note: Total cost at $1,228 per day = $23,580. (All costs are reported in Canadian dollars.) Total of 144 hours.

lasted one and a half hours, for a total of four hours. All the engineers were present for the entire course. Staff from BIPD assisted the independent course vendor in developing the routing course and cofacilitated the pilot course delivery. Table 2.5 summarizes the cost, the number of hours, and the percentage of the total time for each developer task.

2.9.3 Additional costs

Depending on the circumstances of each organization, costs additional to the ones factored into the analysis above may need to be taken into account. Some of these additional expenses might include the following.

- Technical support staff for administration and maintenance of systems;
- Project management staff;
- Installation fees for software, hardware, and telecommunications connections;
- Ongoing monthly charges for telecommunications;
- Staff and equipment for user help desk;
- Online tutoring for students;
- Expenses associated with customer billing (including charges to business units for the use of services of other business units; e.g., Web hosting of course content on server farm);
- Other costs, such as profit margin that a service provider passes along to internal or external customers.

If expenses such as the ones above are not sunk (i.e., are newly incurred as a result of the Web-based training project), these costs could affect the outcomes of financial analyses such as break-even and ROI. However, in many cases, additional costs such as these are small. For example, in many organizations, one administrator manages several servers, currently available bandwidth is sufficient to accommodate Web-based training delivery, and current staffing levels are adequate to provide project and user support.

2.10 Ratio analysis

2.10.1 Break-even number of students

The break-even number of students for each course is listed below.

- Frame relay course: 51 students ($88,364−28,600)/($1,284−110);
- TCP/IP course on WebCT: 111 students ($158,708−28,600)/($1,284−110);
- TCP/IP course on Mentys: 112 students ($160,049−28,600)/($1,284−110);
- Routing course: 4 students ($28,600−24,332)/($1,284−242).

2.10.2 Return on investment

2.10.2.1 Expected number of students per year

To calculate ROI, an assumption must be made about how many students will be trained per year with a course. The number of students trained per year is used to calculate the fixed costs per student so that these costs can be amortized over all students who will use the course. Otherwise, we would be trying to recover all fixed costs with the first student, which clearly cannot be done and which fails to take into consideration the reusability of both the course itself and the hardware and software used to deliver it. In the present study, it was expected that 30–40% of Bell Canada's 300 wide area network (WAN) engineers would take this course, equaling 125 participants. Another 75 participants from other Bell Canada departments were also expected to take each of these courses. Thus, the expected number of students per year in this case was 200 students per course.

2.10.2.2 Platform costs per student

The platform costs per student (rounded) are as follows:

- WebCT: $0.14 per student ($27/200);
- Mentys: $7 per student ($1,368/200);
- Pebblesoft: $8 per student ($1,596/200);

- Symposium: $3.50 per student ($684/200).

2.10.2.3 Salary costs per student for classroom training

To calculate ROI it is necessary to know the savings per student for Web-based delivery over classroom delivery. Therefore, the delivery cost per student in a classroom must be determined. The cost of offering these three courses in a classroom to 200 participants per course annually would be $122,769 per course or $368,308 for all three courses. Assuming a classroom course would be offered to 200 students per year, the classroom course development costs per student would equal $143 ($28,600/200). Table 2.6 presents an analysis of the

Table 2.6

Savings per Student

Classroom Course Costs				
Course development			$143	
Tuition			$600	
Travel			$70	
Salary of students			$614	
Total			$1,427	
Web-Based Course Costs				
Item	**TCP/IP Course on WebCT**	**TCP/IP Course on Mentys**	**Frame Relay Course on Pebblesoft**	**Routing Course on Symposium**
Course development	$660	$660	$434	$53
Server ($15,000/3 courses/200)	$25	$25	$25	$25
Learning platform	$0.14	$7	$8	$3.50
Salary of students	$110	$110	$110	$176
Salary of instructor	—	—	—	$66
Total	$795	$802	$577	$324
Savings per student over classroom delivery	$632	$625	$850	$1,103

savings per student for each course. The ROI for each course in the expected case of 200 students trained per course per year is found in Table 2.7.

The present value of the cost savings was estimated to be 3.621, an amount that takes into account the five-year life span of the courses and a discount rate that reflects a moderate investment risk. The ROI was positive in all cases, although the courses with the least amount of multimedia content—the synchronous routing course and the asynchronous frame relay course—had the best ROIs. The TCP/IP courses had the highest percentage of multimedia content as well as the highest break-even points.

Table 2.8 summarizes the number of multimedia development hours per course, the corresponding break-even number of students, and the ROI over five years. The results indicate that all courses will break even within the first year if 200 students per year take each course.

2.11 Discussion

2.11.1 Comparison of multimedia content per course

There was a significant difference in the amount of multimedia content in the courses and, therefore, in the amount of time spent in multimedia development. The synchronous course contained only a few graphics, while the asynchronous courses contained a higher percentage of graphics as well as audio and photographs. The TCP/IP course had the

Table 2.7
Course ROIs

Course	ROI	Dollars Saved for Every $1 Spent
Routing (synchronous)	3,283%	$33
Frame relay (asynchronous)	697%	$7
TCP/IP (asynchronous using WebCT)	228%	$3
TCP/IP (asynchronous using Mentys)	283%	$3

Table 2.8
Multimedia Development Hours, Break-Even Points, and Five-Year ROIs of Courses

Course	Multimedia Development Hours	Break-Even Number of Students	ROI Over 5 Years
Routing	144 hours	4 students	3,283 percent
Frame relay	1,156 hours	51 students	697 percent
TCP/IP on WebCT	1,487 hours	111 students	288 percent
TCP/IP on Mentys	1,487 hours	112 students	283 percent

largest percentage of multimedia because it included a five-minute video segment. The percentage of development time spent in multimedia production and design for each course is listed below.

- Synchronous routing course: 5%;
- Asynchronous frame relay courses: 33%;
- Asynchronous TCP/IP courses: 45%.

2.11.2 Importance of each costing element

While this is just one case study, there still appears to be evidence that the costing elements chosen for it are all important. There was some variation in the use of any particular element among the courses, although the variations were small and some elements were consistently used more or less frequently for all of the courses. None of the costing elements could be discounted as unimportant, and none of the costing elements clearly dominated the others in importance. More research needs to be done in the area of costing analysis for Web-based course design and delivery, but practitioners working in the field may still find the conclusions of this study useful for their own cost-benefit projects.

2.12 Conclusion

Web-based training has higher fixed costs than classroom-based training; however, these higher course development costs are offset by lower variable costs in course delivery. This is primarily due to the reduction in course delivery time (course compression) and the potential to deliver courses to a larger number of students than in traditional classrooms without incurring significant incremental costs. Realizing savings for Web-based courses requires a sufficient number of students to recover course development costs. Since employees must be paid for time they spend taking a course, student salaries are an important consideration in this costing study.

While all of the measures of financial performance indicated that there is a strong business case for Web-based training, financial indicators suggest that the amount of multimedia content in a course is the most significant factor in cost. The TCP/IP course, which had the least cost savings and a break-even number of students of 112, contained a five-minute video segment. As a result, the number of hours spent in multimedia production was far higher than for the other courses. The synchronous routing course, which was the most cost-effective course in the pilot with a break-even number of four students, contained only a few graphics and live audio. The limited amount of multimedia content in the synchronous course offset the higher costs of course delivery resulting from the cost of having a live instructor present during delivery and the greater student salary costs due to the extra time required to deliver the course.

In comparing the number of hours spent on multimedia development across all courses, the variations are relatively small. This indicates that each of the elements discussed previously are important to consider when costing Web-based course development. The methodology used in this case study, therefore, can be used in future cost-benefit studies of Web-based training.

References

[1] Russell, Thomas, *The "No Significant Difference" Phenomenon as Reported in 248 Research Reports, Summaries, and Papers,* 4th ed., Raleigh, NC: North Carolina State University, 1998.

[2] Aubé, Stephan, Michael Chilibeck, and David Wright, *Cost Analysis of Video Conferencing for Distance Education at the University of Ottawa*, Working Paper; 96–45 [Ottawa, Canada]: Faculty of Administration, University of Ottawa, 1996.

[3] Trevor-Deutsch, Lawry, and Walter Baker, *Cost/Benefit Review of the Interactive Learning Connection, University Space Network Pilot*, Ottawa, Canada: Strathmere Associates International Ltd., June 1997.

[4] Hall, Brandon, "The Cost of Custom WBT," *Inside Technology Training*, July/August 1998, pp. 46–47.

[5] Bates, A.W., *Technology, Open Learning and Distance Education,* London: Routledge, 1995.

[6] Bartolic-Zlomislic, Silvia, and Tony Bates, *Assessing the Costs and Benefits of Telelearning: A Case Study From the University of British Columbia*, Vancouver: University of British Columbia, 1999, (http://research.cstudies.ubc.ca/).

[7] Higher Education Information Resources Alliance, *What Presidents Need to Know About the Payoff on the Information Technology Investment,* Background Papers for HEIRAlliance Executive Strategies Report #4, Boulder, CO: CAUSE, 1994.

[8] American Society for Training and Development (ASTD*), Bibliography: Return-on-investment*, Alexandria, VA: ASTD, 1998 (http://www.astd.org/); American Society for Training and Development (ASTD), *Bibliography: Evaluation*, Alexandria, VA: ASTD, 1998 (http://www.astd.org/).

[9] Phillips, Jack J., "How Much Is the Training Worth?" *Training and Development*, Vol. 50, No. 4, April 1996, pp. 20–24; Phillips, Jack J., "ROI: The Search for Best Practices," *Training and Development*, Vol. 50, No. 2, February 1996, pp. 42–47; Phillips, Jack J. "Was It the Training?" *Training and Development*, Vol. 50, No. 3, March 1996, pp. 28–32.

[10] Hall, Brandon, *Web-Based Training: A Cookbook,* New York: Wiley, 1997.

3

Videoconferencing Compared With Web-Based Training Delivery

3.1 Introduction

Distance learning, which uses the Internet or another network to deliver learning material, is an example of electronic service delivery. The Web and videoconferencing are the fastest-growing modes of distance learning in North America. This chapter provides a brief overview of videoconferencing and compares the costs of using each of these technologies for distance-learning applications. While this book primarily deals with Web-based training, the comparison with videoconferencing helps illustrate situations in which Web-based training would be most appropriate. This chapter also discusses the emergence of collaborative learning over networks and provides examples are provided of how organizations are implementing this new learning paradigm.

3.2 Videoconferencing

Videoconferencing allows a teacher and students at different geographic locations to communicate using both audio and video and, usually, text and graphics. Because all participants are present at the same time, this type of technology-based communication is referred to as synchronous. By comparison, with Web-based learning, students access the learning materials over the Web, but the instructor is not usually present within the same time frame. This is known as asynchronous communication.

Videoconferencing using Web servers is becoming increasingly important in new distance-learning implementations. However, the vast majority of installed videoconferencing systems for learning applications use non-Web-based communications. Therefore, in this analysis, we distinguish between synchronous videoconferences and the asynchronous use of teaching material on a Web server, and analyze costs separately.

Most videoconferences are conducted over telephone networks using integrated services digital network (ISDN) or analog lines. Most videoconferences sent over the telephone network are coded at multiples of 64 Kbps. The more multiples purchased, the better the video quality.

3.2.1 The multicast backbone of the Internet (Mbone)

An alternative way of conducting videoconferencing is over the Internet using the Mbone, the multicast backbone of the Internet. Through multicasting, course information such as lectures, teaching materials, and interactions between instructors and students can be sent and received by students at multiple destinations. However, the Mbone is currently used most often for research seminars rather than for mainstream distance-learning applications. Figure 3.1 illustrates how the

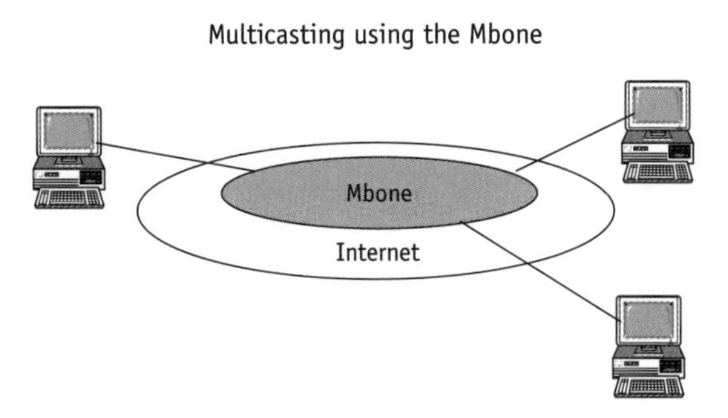

Figure 3.1 Multicasting using the Mbone.

Mbone multicasts audio, video, and data traffic from a teacher at one site to all the sites that have asked to receive a session.

Copies of Mbone traffic are made at the latest possible point on the Internet, thus eliminating the need for the source location to send out multiple copies of data streams. Users' desktop computers are used to play the audio and display the video and data, and the end user controls how the information is displayed. For example, a student could choose to display a small video window that shows the head and shoulders of the teacher and a larger window for the teaching materials. When a student asks a question during an Mbone session, the audio, video, and data from that student are multicast both to the teacher and all participants in the multicast. In this way, a "class" discussion can be held.

Today, the Mbone is available only on certain parts of the Internet, primarily the Internet backbone. For distance learning and other applications, users must go to a location with a high-speed Internet connection to view an Mbone session, and all participating organizations need to have a direct link into the Internet backbone. Although this technology is not commonly used today, whenever ISPs choose to implement the Mbone—which technically amounts to a software upgrade on their routing equipment—it will become a very efficient way of delivering motion video, audio, and graphics multicasts to students' homes.

3.2.2 Costing analysis for videoconferencing

Videoconferencing is used in learning applications for several reasons:

- To provide education and training services to students in remote locations that do not have ready access to qualified teachers;
- To reduce costs associated with travel and instructors' salaries when students in multiple locations need to take the same course;
- To expand a teaching institution's current market for students by reaching a wider audience.

Videoconferencing is used for learning applications by academic institutions, government, and corporations, and by schools teaching K–12 students. M.B.A. programs have been longtime users of videoconferencing to reach remote locations. Queen's University in Kingston, Ontario, has offered students this option for its executive M.B.A. program since 1994, and the videoconferencing program now operates in 24 cities. Students enrolled in the program believe that the ability to discuss issues with students from diverse backgrounds living in a wide geographic area is one of the main attractions of this program [1].

Some examples of how organizations are using videoconferencing for training include the U.S. Naval Postgraduate School, which has installed an asynchronous transfer mode (ATM) network to the desktop to connect 2,500 students, teachers, and other users of the system. The network transmits streamed video and videoconferencing [2]. Likewise, the insurance industry in some Canadian provinces such as British Columbia has adopted videoconferencing technology to deliver government-mandated education programs to brokers in outlying geographic regions [3].

Schools are also taking advantage of technology to enhance the learning experience. In San Diego County, 589 schools are able to use five satellite distance-learning classrooms that broadcast classes using microwave transmissions and provide interactive learning using two-way videoconferencing. As well, courses are broadcast over cable channels that reach approximately 850,000 homes [4].

The following costing analysis of the Franco-Ontarian Distance Education Network (FODEN) in Ontario, Canada, is presented as an example of a videoconferencing system currently used in a learning application [5]. Francophone students at the college level are geographically distributed throughout Ontario, and FODEN is used to deliver courses in French to locations where it is not cost-effective to employ a Francophone teacher. The analysis presented here forms the basis for economic decisions used by FODEN administrators concerning when the system should or should not be used to deliver courses.

For simplicity, this analysis assumes that two sites are linked. However, more than just two can be linked, and, in fact, cost-effectiveness improves by adding more sites. Instead of employing two professors at two universities to teach the same course to students, one professor is employed at one university, and classes are videoconferenced to the other university. The videoconferences are delivered to university classrooms rather than to students' homes. Thus, the students still need to come in to the university campus for classes. Figure 3.2 presents a cost comparison for a three-credit, 39-hour university course between two sites, first showing the videoconference cost and then the regular classroom cost.

Variable Costs for Videoconferencing

- 3 Credit Course: 2 Sites
  - One Professor $7,000
  - Telecommunications $1,300
  - Honorarium $ 400
  - Course materials/shipping $ 100
  - TOTAL $8,800
- Regular Classroom
  - Two Professors $14,000
- **Savings** **$5,200**

Cost per course

Figure 3.2 Variable costs for videoconferencing.

The cost of the professor's time is, on average, $7,000 per course, and the telecommunications cost is about $1,300 (all costs are reported in Canadian dollars). Although the professor gets paid a small honorarium for participating in a videoconferencing session, he or she does not get paid more for any supplementary work involved to convert course material into a format suitable for a videoconference. In addition, there is a small cost for shipping hard copy materials to the other site. The total cost per course is approximately $8,800. On the other hand, if each university offered the course in a traditional classroom setting, the cost would be $14,000 to employ two professors. Therefore, the savings per course amounts to $5,200 for using videoconferencing rather than traditional delivery.

Figure 3.3 lists the annual fixed costs for the system. The cost of adapting the classrooms for videoconferencing and for the equipment in the rooms is $56,000 for two high-quality videoconferencing systems. There are also associated salaries for the staff employed in the distance-learning departments to maintain the equipment. A video bridge is used when communication is made between more than two sites, although it is possible to lease capacity on video bridges through a telephone company. However, multisite conferencing is used in FODEN frequently enough for an on-site video bridge to be more economical. Graphics bridges are another component of the system, and

Fixed costs for videoconferencing

Annual fixed costs	
2 Videoconferencing systems	$56,000
Salaries	$70,000
Video bridge	$18,000
Telecommunications	$10,000
Graphics bridge	$ 1,500
Total	$155,500
	Cost per system

Figure 3.3 Fixed costs for videoconferencing.

there is also a fixed telecommunications cost for leased lines, independent of any courses taking place. This telecommunications cost is in addition to the telecommunications cost in Figure 3.2. The total capital cost is equal to $155,500 for a two-site videoconferencing system.

3.2.2.1 Break-even analysis

The break-even point can be calculated from the fixed cost of $155,500 and the savings per course of $5,200, giving a break-even point at approximately 30 courses, as illustrated in Figure 3.4. In other words, if this system is to deliver at least 30 courses between two sites, then there is a savings.

3.2.3 Delivery issues

There are a number of key issues in the delivery of courses using a videoconferencing system.

3.2.3.1 Number of sites

The break-even can be improved and the system made more efficient by delivering courses to more than one remote site. In other words, three or four sites will be more cost-effective than one or two.

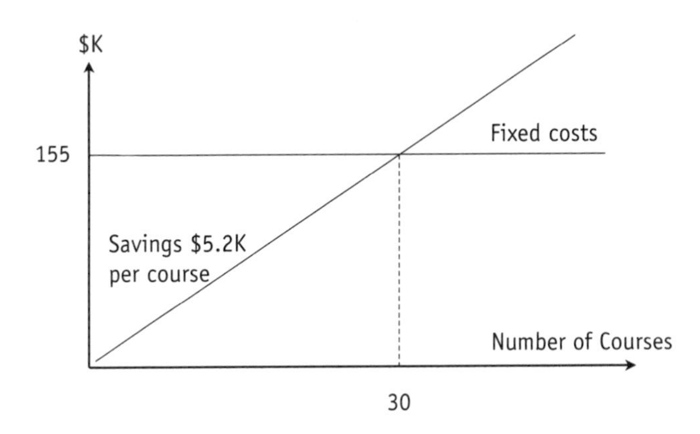

Figure 3.4 Break-even analysis for videoconferencing.

3.2.3.2 Video quality

Section 3.2.2.1's analysis is done at the minimum video quality typically used for distance learning, which is two multiples of 64 Kbps. If we pay for more than two multiples of 64 Kbps, then there is an additional cost per course, but the video quality is improved. How good the video quality needs to be depends on the environment of the educational institution. In Ontario, education is not a competitive commodity as it is in some parts of the United States. In California, for instance, distance education is done at the highest possible video quality because universities and campuses are competing with each other, often setting up distance-learning sites in the cities of the other universities. Therefore, it is important for the video quality to be comparable to viewing a live professor in a traditional classroom. In the less competitive environment of Ontario, there is a lower cost associated with video quality. Training, on the other hand, is a competitive market, and video quality is important to achieve student satisfaction and market share.

3.2.3.3 Delivery to students' homes

Delivering education to students' homes costs significantly more per student if we use the telephone company network for videoconferencing at multiples of 64 Kbps. The main cost associated with delivering it to the student's home is the cost of video bridging, where there is a cost for each connection to each home. Delivery can be done at a much lower cost over the Internet using the Mbone, but Mbone service is not generally available to residential users from most ISPs. Of course, many university courses are broadcast to students' homes using cable TV, another competing technology. The advantage of videoconferencing compared with cable TV is that videoconferencing is interactive (i.e., students can ask questions).

3.2.3.4 Administrative use of facilities

Organizations can also use videoconferencing networks for meetings and other administrative purposes in addition to teaching, thereby saving money in ways not related to learning and improving the overall business case.

3.2.3.5 Cost of adapting materials

In the case of university courses, the cost of adapting teaching materials for the videoconferencing environment is minimal, since the work is generally regarded as part of the professor's job, and no additional payment is required. In a corporate training environment, there could be a more tangible direct cost to the corporate training department for adaptation if specialized services were employed to ensure that presentation materials were of professional broadcast quality.

3.2.4 University teaching versus professional training

In the case of professional training, a company frequently hires a third-party trainer who may charge a higher fee if the sessions are multicast to different locations. In the case of university teaching, since the professor is an employee of the university, he or she gets only a small honorarium over and above his or her salary if a session is multicast. An Internet discussion group, the Distance Education Online Symposium (DEOS-L), has surveyed honoraria paid to professors and found that it ranges from zero to about $1,000 per course. This is a minimal amount compared with the amount a third-party trainer may negotiate for a distance-learning course.

Another point about the commercial situation is that a multisite corporation can use its private network for videoconferencing, whereas a university typically needs to lease additional facilities from a telephone company. The cost of bandwidth is, therefore, often lower for a corporation than for an educational establishment. In summary, the commercial situation is different from the university situation in that the cost of the trainer may be higher and the cost of the networking may be lower.

3.3 Web-based training

The main cost difference between videoconferencing and Web-based training is that there is a large cost in adapting the teaching materials for a Web server, whereas, in the videoconferencing situation, the teaching materials do not need to be adapted as much. Figure 3.5

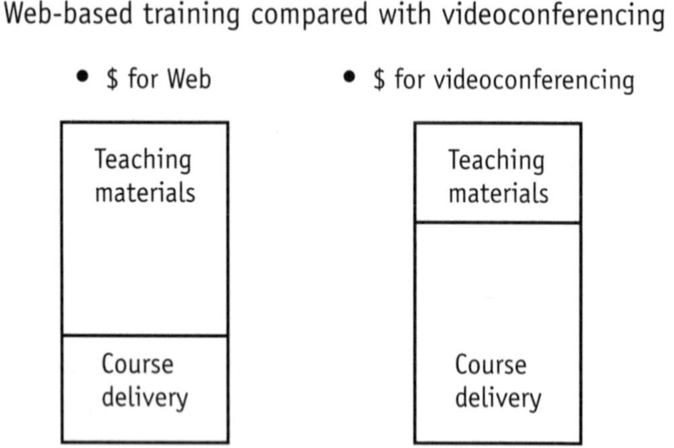

Figure 3.5 Web-based training compared to videoconferencing.

illustrates the comparative costs of Web-based and videoconferencing delivery.

The reason that materials need to be adapted more for Web-based training is that the materials must be essentially self-teaching. Therefore, the teaching materials need to be much more self-explanatory, with a lot of integrated learning exercises. For this reason, instructional design is much more important in a Web-based course than in a videoconference-based course.

On the other hand, the savings in Web-based training come at the time when the course is delivered, due to savings in instructor salaries. In the case of corporate training, students are employees, and the employer is paying them a salary for the time they are participating in a course. The more quickly the students can get through the course, the less the cost to the employer. It has been found that students cover Web-based training materials much more quickly than they do in a live session in a classroom or in a videoconference. Students are able to skip material that they already know or material that is irrelevant to their job. The Web-based course compression ratio is between 20% and 80% for students' time compared with a classroom or videoconference course. In other words, the amount of time the students take to do a Web-based course is between 20% and 80% of the time that they would have taken to learn the same material in a classroom or videoconference delivery format. The savings accrued from students' salaries is significant. Course compression is not an advantage in the university situation because universities do not pay students for the time they spend taking courses.

3.3.1 Costing analysis for Web-based training

Chapter 2 provides a detailed cost analysis of Web-based training. This chapter presents an overview of the major factors entering into that analysis in a format that can be compared with the videoconferencing situation. As with the videoconferencing situation, we present variable and fixed costs and how they can be combined in a break-even analysis. Figure 3.6 gives the variable costing elements for delivering a Web-based course.

It is important to note that the costs listed in Figure 3.6 are *per student*, whereas the costs in Figure 3.2 are *per course*. When we deliver a course from a Web site, the first cost element is the student's time. There is also a telecommunications cost, which is usually small since delivery is over the Internet. Finally, there is the instructor's and students' time for whatever interaction is built into the instructional design. For instance, a minimal level of interaction would be to allow

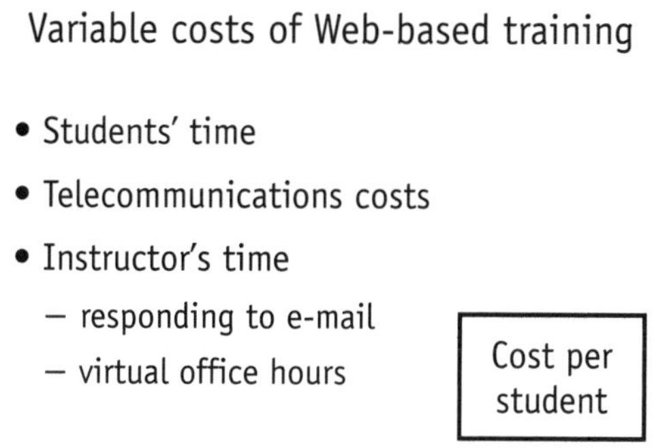

Variable costs of Web-based training

- Students' time
- Telecommunications costs
- Instructor's time
  - responding to e-mail
  - virtual office hours

Cost per student

Figure 3.6 Variable costs of Web-based training.

students to ask questions via fax or e-mail. At the other end of the range, some instructional designs include virtual office hours, whereby students can set up multimedia sessions with the instructor for more extensive discussions.

The fixed cost elements are indicated in Figure 3.7. It is important to note that these costs are *per course*, whereas the costs in Figure 3.3 are *per system*. The instructional design and the multimedia design costs are the main differences from classroom-based training. There is a high cost associated with graphics or video content in the teaching materials, and other cost elements are indicated in Figure 3.7.

3.3.1.1 Break-even analysis

The break-even point, where the cost of the Web-based training matches the cost of classroom-based training, is indicated in Figure 3.8. Initially the cost is higher because of the cost of putting the material on the Web, but the cost per student of delivering the material is lower than in the classroom. The classroom cost is a step function because of the pedagogical reasons that limit the maximum number of students that can be accommodated in a single classroom with one instructor.

In Figure 3.8, the horizontal axis is the number of students, whereas in Figure 3.4 it is the number of courses. In Web-based training, the break-even analysis indicates the number of students required to make a course economical. In videoconferencing, because of the

Fixed costs for Web-based training

- Instructional design
- Multimedia design
- Text/audio/graphics/photo production
- Software development
- Content integration
- Testing
- Editing

Cost per course

Figure 3.7 Fixed costs for Web-based training.

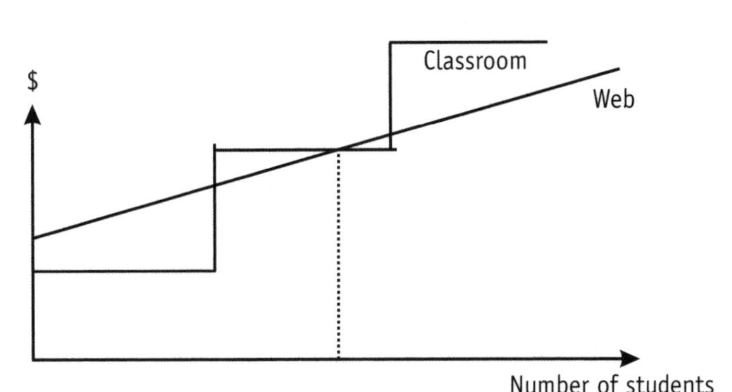

Figure 3.8 Break-even analysis for Web-based training.

much higher cost of the capital equipment, the break-even analysis addresses the issue of how many courses must be delivered to offset the cost of a videoconferencing system.

3.4 Collaborative learning

Participants in professional training courses often have a rich education and employment background that allows them to contribute meaningfully to course discussions and the creation of supplementary course materials. These interactions can also encourage a sense of community among participants. For example, IBM believes that contact among members of its virtual sales force, who are expected to be the greatest users of its Lotus Notes-based courses, will be one of the greatest benefits of this type of training. While the increase in the virtual workforce has been a great success for IBM in terms of cost benefits and client satisfaction, the isolation of these workers is an enduring management concern. The dropout rate for CBT is high, and one of the reasons for this may be the social isolation of students. Communication via asynchronous text may reduce this isolation and create an atmosphere similar to that of a classroom. As a result of collaboration, there may be mentoring between the instructor and students and among the students themselves.

Both videoconferencing and Web-based training build upon conventional paradigms for learning. Videoconferencing takes a conventional classroom and allows it to be operated at a distance, but the formal roles of student and teacher remain. Web-based training takes the conventional textbook of teaching materials and facilitates interactivity and cross-referencing, but the formal concept of a student accessing teaching materials remains. Client-server architectures make real-time and asynchronous communications between students possible, along with instructor monitoring of these discussions. The text of course discussions and other contributions can be centrally stored and accessed from database files kept on the server computer. Telecommunications technology, in particular the Internet, can be used to enable a new paradigm for learning.

Although student contributions can produce valuable exchanges of information and enhanced course materials, getting busy professionals to take the time to participate in discussions and exercises is difficult. In a college or university setting where students take courses for credit, professors can require that students post messages to discussion groups and submit assignments to receive a passing grade. Instructors of professional training courses rarely have the same authority over their students, and getting voluntary participation is a greater challenge. However, when a course is related to certification, such as the MOCOMP program for physicians described below, some of the problems with motivating students to participate are solved.

The Royal College of Physicians and Surgeons (RCPS) in Ontario operates the Maintenance of Competence Project (MOCOMP). It allows doctors, particularly family doctors, to share their problems, the types of cases they are encountering, and their experience on how to treat unusual cases. MOCOMP is essentially an Internet discussion group among professionals. While it is not unusual for a group of professionals to have a discussion among themselves, MOCOMP is different because RCPS monitors the archive of discussion and *certifies* that the doctor has participated in the program. MOCOMP is part of the doctors' ongoing training after they are licensed to practice medicine, ensuring that family doctors won't be isolated from new medical developments. This method of learning through the discussion of issues where there is no formal teacher and no formal students results in a new paradigm for learning where every participant is a peer.

Doctors participating in MOCOMP maintain a learning portfolio in a PC-based diary and submit and respond to "critical-thinking" questions in an electronic Question Library. Taken together, these constitute a record of doctors' participation in MOCOMP and can be compared with an artist's portfolio of work. The learning portfolio supplements a doctor's personal résumé as a record of workplace training. The certificates issued by RCPS indicate doctors' contributions to group discussions and their self-directed learning.

Requiring participation in discussion groups also means that a certain amount of flexibility in terms of when a course can be taken—one of the greatest strengths of distance learning—is lost, because regular participation in discussion groups is necessary for the groups to work well. Students cannot stop a course partway through and resume it at a later date, because the momentum of the discussion would be lost. Real-time discussion groups such as those using videoconferencing are even more inflexible than asynchronous groups, since participants all need to be present at the same time. When students do participate in discussions and exercises, course designers must decide how this material is best incorporated into course design. One solution is to create a database and text files to store this information and make the files available to future course participants. This was done by the designers of the "Roadmap to ATM" course at Algonquin College.

The Roadmap to ATM is a collaborative project funded by Algonquin College, the Communications Research Centre, Knowledge Connections Corporation, and the Ottawa Carleton Research Institute Network. Algonquin College took responsibility for the design and implementation of the course. The course was first offered in 1997 to professionals working in broadband-related fields in business and government in the Ottawa-Carleton region. The goal of the Roadmap to ATM project is to deliver training on the subject of ATM technology; the program uses ATM as the method of course delivery. ATM was chosen for course delivery because broadband capabilities allow rich multimedia content and a wide choice of communications, from e-mail to videoconferencing. High-speed broadband communications like ATM reduce wait times in sending and receiving information. When communication and database use is less frustrating, participation is likely to be more frequent. The course designers wanted to incorporate what they learned from participants each time the course was offered in this fast-changing subject area. Incorporating student participation into the course design was seen as a cost-effective solution (Meike Miller, personal communication, 1997).

Academic institutions are connecting students and teachers to online resources and one another, resulting in a wide variety of learning interactions. For example, students at the Harvard Business School use a $10 million campus-wide intranet to communicate with professors, conduct research, prepare class assignments, and participate in collaborative study groups [6].

3.4.1 Systems for Web-based training

Software systems for the authoring and delivery of Web-based courses act as templates in the design and delivery of Web-based courses and are the building blocks for this mode of training. They determine issues such as the way in which a course is structured and designed, the animation of learning sequences, the approach to test items, and the type of interaction the learner can have with course materials. The increasing power and flexibility of computer technologies and software requires increasingly complex authoring tools that can translate these advances into applications.

The client-server model allows for more elaborate services than simply presenting lessons and examinations to the students. Because multiple clients may connect to the server, the server may accept messages from one student and send a reply to another student. One use of a common gateway interface (CGI) binary script is to act as a reflector when two or more users are logged on to the server at the same time and request communication (e.g., when multiple sockets are established). A socket is part of a computer operating system that allows one computer program to send a message to another program; the programs each may be running on separate computers. There can be communications between students, either in real time, or asynchronously through a Usenet-like facility. These communications can take the form of "class discussions," casual "hallway conversations," or "tutorials" with the instructor.

The way the Web implements a standard client-server model is sufficient for asynchronous, Usenet-like applications because the server can execute CGI programs to store messages from one student in its files and include those messages in its replies to other students. However, typical client-server architectures are inadequate to support real-time interactions between students, which would resemble computer-based conferencing, because sockets are not maintained that persist longer than the duration of a single request and response. The Web server does not track sequences of interactions with a single client such as would be required to maintain a dialogue between people.

Web browsers have an excellent capability to present multimedia information to students. They use the Internet-addressing protocols to find the correct server. A browser translates a click on a hypertext link into an Internet address and file name, opens a socket to the correct server, requests the file, receives the file, and then closes the socket. This process employs Java applets, small computer programs written in the Java programming language that are specifically intended to be executed by Internet browsers. A distance-education course on the Web may have lessons on more than one server.

The Web is based on temporary connections, and neither the client program not the server has any concept of a session or a user during a session. Servers view each individual request as a separate and independent interaction. If a client requests a number of Web pages from the same client in a row, each request is independent of the others. Each request opens a new socket and closes it when the request is satisfied. The most recent developments in Web client-server programs pertain to the ability to maintain persistent sockets, but most Web servers today do not yet allow this.

Despite this, users may think of themselves as "logged on" to the server. Conceptually, a student who is communicating with the instructor and other students via text messages can feel as though these interactions are in many ways similar to face-to-face communication.

Figure 3.9 illustrates how communication is achieved between the clients and server and which types of files are used in collaborative learning.

Several new tools in the marketplace give course producers, instructors, and users the functionality to create new learning experiences. For example, Lotus Development Corp.'s (www.lotus.com) Lotus LearningSpace uses a Web browser to allow instructors to customize courses through a development tool, and students are able to use the application to view live training sessions and collaborate on course assignments. Training management features include monitoring results from the classes and saving course assignments [7]. Toyota's Technical Education Network delivers online training materials to 50 training facilities across the United States and has training management features including student registration and tracking [8].

Centra Symposium is a tool that allows instructors and students to participate in live audio sessions; visually share course materials, including external applications such as a systems, applications, and

Collaborative learning using Web servers

Client
Internet
Server
1. Get request from user
2. Establish a socket
3. Request CGI script
4. Execute CGI script
5. Get Java applet
6. Return Java applet
7. Disconnect socket
8. Execute Java applet

Java applet
CGI program
9. Establish a socket
10. Information from user
11. Information from user
12. Result
13. Result

Figure 3.9 Collaborative learning using Web servers.

products in data processing (SAP) database; break out into collaborative work groups; and surf the Web. PricewaterhouseCoopers is one large company that uses Symposium for both employee training and as a product offered to its enterprise resource planning (ERP) practice customers [9]. Interestingly, Centra Software (http://www.centra.com) has created a professional certification program in live online collaboration, aimed at training managers, administrators, and users [10].

3.4.2 Video Web hybrids

Web technology is at an early stage of development, and new products that blur the boundaries between audio, video, text, and graphics are being created. One such product is CU-SeeMe Pro 4.0 from White Pine Software Inc. (http://www.wpine.com/cu-pro-beta), which allows for Web-based videoconferencing [11]. Another video Web hybrid is Convergence, a WebDVD platform from Spruce Technologies (http://www.spruce-tech.com) that allows learners to play a digital video disk at their own PC while an instructor at a remote location coordinates text and discussion arriving via the Web with the video being viewed on the local disks [12].

Many organizations have chosen to offer users both videoconferencing and Web-based technologies for collaborative learning applications. For example, the Brookings Institution allows scholars in a wide geographic area to conduct seminars and panel discussions using videoconferencing and Web-based communications [13]. Another organization that has embraced the use of videoconferencing and Web-based courses along with traditional classroom instruction is the National Association of Printers and Lithographers' Center for Professional Development [14]. The recently created Western Governors University also offers students a variety of technologies, including videoconferencing and the Web, to take the courses they broker from colleges, universities, and businesses across the United States [15].

Meanwhile, teleconferencing and videoconferencing companies such as Octave Communications, Contigo Software Inc., and Polycom Inc. are responding to customer demand for innovative and cost-effective communications by supplementing their product offerings with Web-based communication [16]. Collaborative virtual work environments using technologies such as whiteboards, threaded discussions, and Web-based course materials have allowed some companies to reduce training budgets by 30–50% [17]. A San Francisco consulting firm, Collaborative Strategies LLC (http://www.collaborate.com), estimates that the dataconferencing and real-time collaboration market will grow 20% between 1999 and 2004 [18].

3.4.3 Webcasting

Webcasting is another interesting use of Web technology. A Webcast can be created by capturing audio and video images with an inexpensive camcorder attached to a VCR and sending the information to a computer with audio and video capture cards that then send the encoded video and audio signals to a Web server. The audience views the session by connecting to the Web site on the server used for the Webcast. Several organizations have experimented with Webcasting for corporate communications. For example, a Webcast at Bell Canada in November 1999 reached 500 employees in several cities logged on to one of four video streaming servers at 300 Kbps. The use of the servers reduced the bandwidth required to deliver the Webcast to this many people, and costs were minimal because existing equipment was used both for sending and receiving the broadcast, and because Netshow software, which comes free with Windows NT, was used to distribute the Webcast.

While the Bell Webcast was distributed to recipients via unicasting, multicasting will allow Webcasts to make efficient use of bandwidth, greatly reducing costs over those associated with videoconferencing or traditional broadcasting. An example of a Webcast using multicasting to distribute information for training purposes is described by H. Michael Boyd (http://www.pbstbc.com/intel_white_paper.html). A joint effort of Intel and the Public Broadcasting Service's The Business Channel in the United States, this Webcast consisted of a panel discussion among experts while participants viewed the event from their desktops. Interactivity was achieved through online polling and by allowing participants to use online messaging to communicate with the panel members. Multicasting is a built-in feature of the new Internet protocol IPv6, and when it becomes a reliable technology, likely within the first few years of the new millennium, multicasting will be supported by all clients and servers.

3.5 Conclusion

Academic and corporate training situations are very different. Videoconferencing is widely used in academic environments as a cost-effective means of education. The cost of the trainer is a major factor in achieving a good return on investment for videoconferencing in corporate training situations, but if the trainer increases his or her fee as a result of having additional students at remote sites, ROI will be adversely impacted. However, Web-based delivery of corporate training achieves a high ROI, primarily through decreasing the amount of time students spend completing courses, thus saving salary dollars. Academic uses of Web-based training do not benefit from this cost savings, as universities do not pay students for the time they spend in class. Academic institutions are motivated more by an interest in expanding their market—for example, by serving the needs of part-time students who have a schedule that does not allow them to travel regularly to a classroom and who prefer to study at home or at work.

In addition, the Internet opens up opportunities for new modes of training where the traditional formal student and teacher roles of the classroom are redefined. Consumer demand for collaborative communication has motivated the companies that develop the software and hardware that make network-based communications possible to create new products that begin to blur the distinction between videoconferencing and Web-based communication. In the future, distance education courses using Web servers will offer enhanced features for collaborative learning, and take advantage of the real-time communication that is possible by establishing persistent sockets between clients and servers; they will also make greater use of Java applets.

References

[1] Carpenter, Rebecca, "The Anywhere MBA," *Canadian Business*, Vol. 71, No. 17, October 30, 1998, pp. 62–67.

[2] Hamblen, Matt, "Naval School Chooses ATM for Training Net," *Computerworld,* Vol. 33, No. 8, February 22, 1999, p. 69.

[3] White, Doug, "Are Associations Meeting the Educational Needs of Brokers in Outlying Regions?" *Canadian Insurance*, Vol. 104, No. 7, June 1999, p. 30.

[4] Nelson Publishing, "1 Building + 1 Technology = Education for 460,000," *Communications News*, Vol. 35, No. 7, July 1998, p. 60.

[5] Aubé, Stephan, Michael Chilibeck, and David Wright, *Cost Analysis of Video Conferencing for Distance Education at the University of Ottawa*, Working Paper; 96–45 [Ottawa, Canada]: Faculty of Administration, University of Ottawa, 1996.

[6] Santosus, Megan, "Wired Education," *CIO*, Vol. 12, No. 1, October 1, 1998, pp. 52–59.

[7] Mateyaschuk, Jennifer, "Enhanced Net Training," *Informationweek*, No. 742, July 5, 1999, p. 76.

[8] Robinson, Teri, "Toyota Tunes Up Training Program," *Internetweek*, No. 742, November 23, 1998, p. 19.

[9] Darrow, Barbara, "Big Five Firm Enlists Centra," *Computer Reseller News*, No. 826, January 25, 1999, p. 30.

[10] Lakewood Publications Inc., "Collaboration Certification," *Online Learning News*, Vol. 1, No. 43, January 26, 1999.

[11] Lakewood Publications Inc., "Beta Videoconferencing," *Online Learning News*, Vol. 1, No. 48, March 2, 1999.

[12] Lakewood Publications Inc., "Rich Video Web Hybrid," *Online Learning News*, Vol. 2, No. 3, April 20, 1999.

[13] Nelson Publishing, "Cyber Scholars Build Virtual Bridges," *Communications News,* Vol. 36, No. 2, February 1999, pp. 80–83.

[14] Cahners Magazine Division, Reed Publishing, "Learning Leads Relocation," *Graphic Arts Monthly*, Vol. 71, No. 1, January 1999, pp. 82–83.

[15] Stuart, Anne, "Passing Without Distinction," *CIO*, Vol. 12, No. 16, June 1, 1999, p. 14.

[16] Securities Data Publishing, "The Trend Is Clear and Audible: Tele-Videoconferencing is Going Online," *Investor Relations Business,* February 15, 1999, pp. 8–10.

[17] Greengard, Samuel, "Web-Based Training Yields Maximum Returns," *Workforce*, Vol. 78, No. 2, February 1999, pp. 95–96.

[18] Lakewood Publications Inc., "Collaboration Growth," *Online Learning News*, Vol. 1, No. 47, February 23, 1999.

4

Web-Based Competency and Training Management Systems for Distance Learning[1]

4.1 Introduction

Web-based technologies have created an opportunity for companies to revolutionize the management and delivery of employee training and development. The interest that corporate training departments have in effectively meeting the needs of the end user is nothing new. Trainers have always been committed to delivering high-quality training that has a positive impact on company performance. What is new is

1. Reprinted with the permission of Idea Group Publishing, *Managing Web-Enabled Technologies in Organizations: A Global Perspective,* 2000, pp.191–209.

the precision with which high-powered computers and high-bandwidth networks allow training departments to implement their customer-focused approach. Through the use of competency and training management systems such as the SIGAL system now being piloted at Bell Canada, organizational training plans can be efficiently communicated throughout the organization, training needs can be linked to the performance evaluations of individual employees, and online training materials can be conveniently delivered to employees at their desktops. This chapter discusses the value to companies of using a Web-based system for competency and training management, using Bell Canada as a case study.

4.2 Background

Web-based training has several advantages that have encouraged companies such as Bell Canada to explore the possibilities of its use for employee training. Asynchronous courses, comprised of HTML pages and, in some cases, multimedia elements such as graphics, animation, video, and recorded audio, offer great flexibility in how course materials are both accessed and used. Since asynchronous courses do not have a live instructor, they can be completed by employees at their convenience, at any geographic location with access to a network connection, and at a pace that is set by each individual. Primary consumers of distance learning are salespeople who, more than most workers, spend time away from normal office settings. For example, IBM's sales and marketing force are equipped with mobile computing equipment and use Lotus Notes on a regular basis to take training from hotel rooms and other locations anywhere in the world where they happen to be working [1].

4.3 Issues, controversies, problems

Many organizations, including those in the telecommunications industry, have had to become increasingly competitive in order to survive. Higher levels of competition have resulted in many challenges that must be addressed through training. A "map" of the competencies required by an organization, including the existing competencies, is the first step in providing training that makes a real impact on organizational performance.

The determination of which specific competencies are required by individual employees is generally developed through continuous communication between employees and managers, keeping in mind the strategic direction of the organization. Dove [2] identifies the need for an organization's competency sets to evolve over time, based on the constantly changing business environment. After competency assessments are complete, appropriate training is required to fill the identified gaps. To be effective, training must be relevant to current (and forecast) needs, readily available to employees, interfere as little as possible with daily work, and result in real improvements in overall productivity.

4.3.1 Managing employee training to support organizational goals

Competency modeling, increasingly used by companies to build performance management strategies, is essentially a detailed description of the skills and knowledge required to perform a specific job. Competency profiles form the basis of both employee evaluations and the creation of individual training plans. On a company-wide basis, competency models map out the relationship between the company's business plan and the aggregate employee skill sets required to implement those business plans.

In [3], Zemke and Zemke describe how companies are using competency profiles to achieve mission-critical goals as a:

- Vehicle to communicate to employees the behavior required to succeed in their jobs;
- Part of the process to create a training and development plan for each employee;
- Method of helping select employees for new roles in the organization.

First, competency profiles provide a clear description of what good performance looks like (describing actions, for example, such as “recognizes when it is important to act even when there is a degree of uncertainty”). Communication of management expectations is particularly important in companies experiencing downsizing or reorganization, where a large number of employees are expected to adapt quickly to new roles in the organization.

Second, a training and development plan must be created to address each employee’s identified needs. Gaps in skills and knowledge are often identified through the use of what are known as 360-degree assessments. In a 360-degree assessment, each employee is given feedback from their supervisor and several coworkers whom they have chosen to evaluate their performance. The criteria for evaluation are based on a specific competency profile for the employee’s job.

Apart from the need for clear performance standards, companies must also have policies and practices in place that support the efforts of employees. Ensuring that reward systems are established encourages the acquisition of new skills and peak performance. A properly designed and supported training management system gives all employees the opportunity to succeed. To be successful, competency-based training must be accompanied by appropriate employee incentives to complete the training. Rubino [4] lists five criteria that must be satisfied for reward systems to be successfully implemented:

1. Successful fit with the organizational environment;
2. A system that is fair to employees;
3. A system that yields financial returns to employees;
4. A system that involves employees and managers;
5. Goals that are made clear through communication.

Tracking training results means that those who have the highest levels of success will have the best chance at promotions, performance bonuses, or other rewards (for example, the opportunity to attend conferences). Training management systems also function as a course catalog and front end for delivering available courses. The gaps that are identified by the 360-degree assessments form the basis for the employee's training plan. Ideally, an employee can access the recommended courses listed in his or her training profile simply by using a computer mouse to click on a course title; the training system will then load the course material ready for use.

Delivery of courses to the desktop falls into the EPSS category of distance learning. EPSS is a term often used to describe "just-when-needed" training that supports a specific job requirement. As Malcolm [5] points out, EPSS's real breakthrough is not so much in the use of technology but in new thinking on the part of training managers. Trainers are beginning to realize that increases in employee productivity can be achieved by providing training relevant to specific job tasks, at an employee's desktop, in an electronic format.

By communicating the criteria for promotion to all employees, Zemke and Zemke point to examples where companies have decreased employee turnover. Essentially, employees perceive that companies are using fair criteria when assessing them for promotion, and this knowledge gives each employee the opportunity to prepare him or herself for promotion through training and development activities. The flip side of 360-degree assessment is that the company is able to obtain valuable feedback on overall company performance, which amounts to an aggregation of the behavior of all employees.

Organizations that manage training are making strategic decisions about their training investments. These organizations operate with more complete information and can make better decisions about how training dollars are invested, as well as ensure that their demand for particular employee skills matches supply. At a local management level, managers are better informed about the competencies of their employees and better able to match employees to the work that needs to be done on a daily basis.

Knowledge of organizational competencies can equally lead to a decision to reduce the level of particular skills that are no longer required or available in overabundance. Although downsizing has earned a poor reputation through ill-considered and indiscriminate implementations, the regular adjustment of staffing levels is essential for companies to remain competitive. Managing training is the first step in reducing layoffs by proactively retraining employees who possess skills no longer required and redeploying employees into new positions.

The third use of competency profiles is to improve the process of filling job vacancies. This is done by developing profiles that can be used as job descriptions when hiring to fill job vacancies, using the profiles to tailor interview questions for job applicants and to describe the skills of employees already working in the company to identify potential internal job candidates. Employers are able to improve the supply of required employee skills by offering training to increase certain skills in the organization, and employees reap rewards through increasing their job opportunities and earnings.

4.3.2 Pay for performance

Directly linking pay to competencies is a rising trend in organizations. Glaxo Wellcome and Guinness both use generic competency profiles to establish pay ranges [6]. While ensuring that employees attain required levels of knowledge is the first step in measuring the success of training, improving job performance is the ultimate goal. Increases in competence are rewarded only when they lead to higher levels of contribution.

Crew [7] argues that a performance-driven system promotes competition, maintains quality, and emphasizes a consistently high level of achievement. The goal of a performance-driven system is to set clear standards and to align resources, policies, and practices with the support that students need to meet the goals that have been set for them. Management principles such as defining clear standards, setting training strategies designed to enable all students to meet the standards, and tracking results are required at all levels of the organization to make performance-driven systems a success.

As Davis, Lucas, and Marcotte [8] put it, the goal of competency-based profiling and assessment is to make a positive impact on the goals, often financial, of the organization through making the best use of human resources. In other words, training is an investment in human resources, and if well-managed, should give a positive return. General Motors is another example of a corporation that is using a combined approach to competency assessment and training management to achieve corporate business goals. General Motors has assessed the desired personal leadership characteristics, business direction, communication, and program implementation needed by managers and linked these outcomes to specific training. To measure the success of the training initiative, that company simultaneously measures changes both in organizational culture and individual performance. Owens-Corning Fiberglass Corp. has taken a similar approach using the Web for training delivery [9]. Employees at Owens-Corning are able to access Web-based training materials that have been identified as key to their career development from training providers such as the American Management Association.

Improvements in performance can be measured in various ways. For example, higher productivity, higher quality output, fewer errors, faster product delivery to customers, or increased sales all would indicate an improvement in employee performance. Performance appraisal systems are the method typically used to identify training requirements and assess training results. Goals are often set both for individual employees and for group effort. For example, an individual may be required to increase his or her level of productivity during a particular period, and a productivity goal will also be set for the work group as a whole.

While some companies choose to link training and performance to direct financial rewards, other organizations simply rely on the power of the performance appraisal system itself to provide incentives for employees to successfully complete required training. If an organization requires that recent performance appraisals be reviewed before employees can receive promotions, there is an indirect link between the successful completion of training and financial compensation. Furthermore, depending on the nature of the organization and the job an employee performs, unsatisfactory performance appraisals may ultimately result in an employee losing his or her job.

Many companies are now using a performance-based pay structure under the theory that employees work harder when their paychecks are at risk. Under this type of system, pay may be treated as follows.

- Entirely based on performance (e.g., salespeople who get paid on commission);
- Partially based on performance (e.g., the top 10–30% of salary is paid only if certain performance criteria are met);
- Nominally based on performance (e.g., base salary is guaranteed, but a bonus is paid if performance meets or exceeds set criteria).

Will employees be able to withstand the pressures that pay for performance brings on a long-term basis? While there are companies that reward above-average performers with pay increases, some pay-for-performance programs are more oriented toward punishing below-average performers. Although the question cannot yet be answered with any accuracy, it seems likely that constant threats to withhold salary or lay off personnel are not sustainable strategies for companies in the long term. Apart from causing employees to continuously seek out more stable employment that then results in costly turnover, inspiring fear in employees seems an unlikely plan for increasing corporate profits.

4.4 Solutions and recommendations

4.4.1 Enterprise-wide competency and training management systems

Competency and training management systems are a key part of managing a company's human resources. Systems such as SIGAL can be viewed as management support systems that help manage organizational knowledge and the skills of individuals through training. Web-based learning platforms, the software systems that house and deliver distance-training courses over the Web, are becoming increasingly sophisticated in features and functionality. Until recently, most of these systems offered little in the way of student administration and training management. However, developers are now paying more attention to the elements that are required to effectively deliver training over the Web, such as the following.

- Administration and management of company training objectives;
- Online employee competency assessment tools;
- Searchable database of collective competencies within the organization;
- Tracking of employee requirements for specific courses through user profiles;
- Online course registration;
- User interface for employees to access recommended Web-based courses;
- Online testing and database of test scores available to students, managers, and administrators;
- Tracking of training budget commitments and expenditures;
- Gateway to broader human resources management systems such as PeopleSoft for linking HR records to employee training requirements and test scores.

The software systems that house and deliver Web-based training content have only recently begun to offer training management and student administration features. In [10], Stamps points out that there are two aspects to enterprise-wide training systems. The first is the ability to deliver training content over networks to employees throughout the organization. The second feature is the ability to manage the training for the entire organization through the use of a central training management system. While organizations have been adopting the use of Web-based and other distributed modes of training in increasing numbers in the past five years, enterprise training management systems are still an emerging area with numerous vendors entering the marketplace with new products.

Phillips and Pulliam [11] identify the seven most important trends in the training and development field as the following.

1. Measuring the effectiveness and efficiency of training and development in a systematic way;
2. Linking organizational and performance needs to program delivery;
3. Shifting the function from traditional training to a performance improvement role;
4. Integrating training into the strategic and operational framework of the organization;
5. Building partnerships with key clients and management groups;
6. Positioning training closer to the work site;
7. Using technology efficiently and effectively.

Vendors have responded to these needs by incorporating student management and course administration features into online course-delivery platforms. Some of the many companies that produce systems with training administration features include WebCT (www.webct.com), Asymetrix' ToolBook II (www.asymetrix.com), Micro Video Learning Systems' Desktop Trainer (www.microvideo.com), ITC Learning (www.itclearning.com), Gyrus Systems (www.gyrus.com), DKSystems (www.dksystems.com), CBM Technologies (www.cbm.com), Criterion (www.criterioninc.com), Live TechSolutions (www.livetechsolutions.com), Pathlore (www.pathlore.com), Pinnacle Multimedia (www.theplm.com), Saba Software (www.sabasoftware.com), Silton-Bookman Systems (www.sbinc.com), Spectrum Human Resource Systems (www.spectrumhr.com), the Desktop Trainer from MicroVideo Learning Systems (www.microvideo.com), and Syscom (www.trainingserver.com) [12].

We have identified six primary features of enterprise-wide competency and training management systems:

1. They house profiles of the skills required by the organization, which form the benchmarks of the key knowledge and skills required for each position.
2. They are a user interface for the competency assessments of each employee, which identify the gaps between current competencies and the ideal competency set.
3. They are the front end for employees to access training that will bridge knowledge gaps, either through direct access to online courses or through providing information in the form of course catalogs for classroom-based training.
4. They enable tracking of whether employees have taken recommended training and whether their posttraining skill levels are sufficient for them to do their jobs.
5. They help manage organizational knowledge, by communicating to employees their expected competencies, competency gaps, and the means of acquiring needed training, and by reporting to supervisors whether employees have successfully completed required training.
6. They provide a vehicle for the ongoing reassessment of the competencies required by the organization.

4.4.2 Organizational requirement for competency and training management systems

While competency management can be done without the use of an online system, the number of details that must be tracked make online systems very effective for that purpose, especially in companies with large numbers of employees. Jones [13] lists distance learning, performance support systems, and training management software among the leading new virtual human resources applications in the knowledge-based organization. Strategies must also be in place to collect, store, analyze, and distribute information and knowledge. As well, a technology infrastructure needs to be in place; change management programs must be implemented to help users adapt to a competency-based learning organization; and the training techniques employed must be results-oriented.

Jones cites Duracell as an example of a company using an enterprise-wide training management system to manage training for its 9,000 employees worldwide. Some of the key features of the chosen system are information management for all training activities, training expense tracking, and an interface to other human resources applications. Duracell evaluated several training management systems on the market against a specific set of needs before choosing a system. Online training management systems are similar to other software products in that each has a unique set of features. Selecting an appropriate system, especially in an emerging area, can present a challenge. To help organizations choose a system that best serves their needs, Brandon Hall [14] has published a report that evaluates the features of a wide variety of online training management products, providing organizations with invaluable information before having to choose a system.

While delivering Web-based courses can provide organizations with clear financial benefits, the use of a training management system to deliver courses does not intrinsically either increase or decrease the costs of training delivery. The real value of using a competency assessment and training management system is less in decreasing the costs of training than in increasing the precision with which training investment can be managed.

4.5 Case description: Use of the SIGAL system at Bell Canada

4.5.1 Competency management at Bell Canada

At Bell Canada, all managers undergo an annual evaluation known as the internal contribution assessment (ICA). Employees are evaluated on their performance of the accountabilities for the review period (achievement capability assessment tool), which is set at the beginning of the year; their leadership capabilities (leadership capability assessment tool); and their technical capabilities (technical capability assessment tool). Capabilities used for the evaluations are included in 18 broad categories ranging from those that span all jobs, such as "general management capabilities," to the job-specific, such as "sales and sales support." These categories are each broken down into subcategories such as "manage projects" or "build and manage customer loyalty."

The subcategories are then even more minutely defined so that a clear picture of each job requirement emerges. For example, under "manage projects," one of the capabilities listed is "understand project management skills and concepts." Altogether, over 500 specific capabilities are defined for potential employee assessment. The next step is for supervisors and peers to rate employees on their performance of each capability using a 360-degree assessment process. Annual employee appraisals for nonmanagement staff are less often done, a problem that the company hopes will be improved with full implementation of the SIGAL system.

The aggregate capabilities of all managers form a picture of Bell Canada's "organizational knowledge," and the resulting database of capabilities serves to help the company understand strengths and weaknesses of the company's management resources. In situations where the company wants to change business direction, this assessment helps senior management make decisions about the skills that need to be acquired through hiring practices, or conversely, to determine which employee skill sets are held in abundance in times of downsizing. The database is also used by human resources professionals for the practical purpose of helping to identify potential job candidates to fill vacant positions within the company.

4.5.2 SIGAL system

The SIGAL system from Technomedia Inc. (www.technomedia.ca) is presently being piloted at Bell Canada for enterprise-wide competency and training management and online training delivery to 200 managers in the network management services (NMS) group described in the case study in Section 4.5.3. The system integrates existing Web-based course content in HTML format, online training delivery through its virtual classroom feature, course management tools, training management data for each employee, and employee records residing within the Peoplesoft human resources management system. SIGAL also does budget tracking for training expenditures.

To use this system, the overall skill set required for the organization is first established by senior management. Training and line managers then translate this information into competency profiles that are the benchmarks for the skill sets required for specific job positions. The next step is to assess how well each individual meets the benchmarks for his or her position. This may be done through employee testing or by asking the line manager to make that assessment.

After employees are assessed, appropriate courses that will fill identified gaps in knowledge or skills are recommended to employees. Web-based courses may be launched directly from the SIGAL system through the virtual classroom features. Postcourse testing determines how well the knowledge has been digested, and when an employee has successfully completed a course, the system then sends an e-mail message with this information to the employee's supervisor.

Web-based training gained the interest of management because of the competitive advantages that just-in-time training brings to the desktop, and for the cost savings over classroom training. Employees involved in various Web-based training initiatives have felt that this mode of training meets their job needs. They have consistently reported that the flexibility of being able to take the courses on their desktops or on their laptops from other locations, flexibility with respect to the time that the courses can be taken, as well as the self-paced aspects of the courses, are all valued features of Web-based training.

Postcourse testing establishes how well an employee has learned the course material, and testing results are stored in the system. Managers may also be asked to assess improvements in job performance. Links with human resources systems such as PeopleSoft enhance SIGAL by providing background information about each employee. For example, the employee's first language can be determined for presentation of the bilingual user interface in the employee's preferred language. Integration with human resources systems allows the management of training to be an integral part of the overall human resources management for the organization. In the future, additional information sharing and reporting between the two systems is likely to occur. Table 4.1 summarizes the features of SIGAL with respect to training management functions and processes.

4.5.3 Broadband training initiative: Case study

The broadband training initiative at Bell Canada is an example of how the company is using the SIGAL system for enterprise-wide Web-based competency and training management. The pilot project consists of 200 Bell Canada managers in the NMS group (Alwyn Mcallen, personal communication, March 9, 1999). The goal of the project is to improve skills in areas critical to job performance, with the added bonus of providing an opportunity for career development. The expected result is highly motivated leaders who will be capable of implementing the company's new strategies in an emerging technological field.

The process begins with employees, supervisors, and coworkers logging on to the system either to complete the 360-degree assessment questionnaire or to access the ICA form itself. All information, including coworkers' 360-degree assessment, supervisor assessments, and recommendations for training, is input online. The result is an efficient and well-managed employee evaluation system.

Part of the pilot includes making 10 new asynchronous courses on the topic of broadband telecommunications available to employees. The courses were identified as "high-need" for the NMS group because the ability to support broadband networks is a new requirement for their jobs. While designing new courses was one of the options considered, existing asynchronous Web-based courses from third parties were found and used in the trial.

Benchmarks for the levels of broadband communications knowledge were set by company training managers, working with line managers in the NMS group. Competency gaps for each employee were also similarly established. After a discussion between the employee and supervisor, the recommended courses are included in the employee's ICA and formally identified as a goal for the next review period. The identified courses are mandatory, and successful completion within a one-year period is required and noted on the ICA for the next review period.

Table 4.1

Training Management Functions, Processes, and SIGAL Features

Training Management Functions	Processes	SIGAL Features
Competency management of organization	Strategic analysis of competency profiles for each business unit and for each position	Searchable database of collective competencies within the organization User interface to external systems
Competency management of employees	Individual competency testing and training plan	Online assessment tools Tracking employee requirements for specific courses User interface to external systems
Course management	Integration with course delivery systems	User interface to access online course catalog User interface to access online courses (i.e., virtual classroom) Online discussion groups for students and instructors Online access to electronic library Scheduling of online and classroom courses including equipment and instructor resources
Student management	Registration, evaluation, tracking test results	Online course registration Database of test scores Gateway to broader human resources management systems
Financial management	Financial tracking of training commitments and expenditures	Financial tracking of training commitments and expenditures
Performance evaluation and economic return	Online student testing Online student course evaluations Tracking test scores and course evaluations Performance appraisals by managers	Integrated online knowledge and application testing Online student course evaluations Reporting features for training management functions Links to human resource management systems such as PeopleSoft

Employees access the courses from their desktops using the SIGAL system as a front-end user interface via the corporate intranet. Two servers are used to deliver the training: one server to house the SIGAL system and the other the training content. During the formal training period, employees start by taking a precourse test to evaluate which, if any, of the 10 Web-based courses is required to fill gaps in their knowledge of the subject area. As employees complete the courses, their supervisors are notified by the system via e-mail, although postcourse test scores are given only to the employees themselves.

Other competency assessment and training management projects are anticipated in the near future with other employee groups. Eventually, Bell expects to use the system to manage competencies and all training, both Web- and classroom-based, for all employees in the company.

4.6 Barriers to the use of Web-based competency assessment and training management

Although related, online competency management and Web-based training delivery are essentially two different functions, each with its own set of problems.

4.6.1 Online training development and delivery

The cost of developing Web-based training courses can be prohibitive. In the BOLI study described in Chapter 2, total fixed costs (in Canadian dollars) for developing Web-based courses were as high as $160,049. Although delivery costs per student are less than those for classroom courses—at Bell Canada, roughly 90% less—organizations need to train a minimum number of students in order to break even. This calculation needs to be made by training departments on a case-by-case basis to ensure that resources are well spent.

Another issue is the availability of adequate computer hardware for employees. More than one organization, including Bell Canada, has investigated the benefits of Web-based training only to find that a large number of the employees need substantial upgrades to the current equipment on their desktops. A related problem is the availability of bandwidth. Even organizations with state-of-the-art intranets may find that existing applications running over the network occupy most of the bandwidth available, leaving expensive upgrades to the network the only option if Web-based training is to be implemented.

Some complex subjects are also difficult for employees to internalize if asynchronous courses are not supplemented with access to a person who can answer student questions. Training departments need to use their considerable expertise in instructional design in order to judge which courses are the best candidates for asynchronous Web-based delivery.

4.6.2 Competency assessment

As Stamps [15] points out, considerable time and effort must go into building the competency models that guide the learning processes in order to maximize the potential of these new products. Like SAP, BAAN, PeopleSoft, and other systems that are market leaders in enterprise-wide systems development, training management systems represent fundamental changes to the way companies conduct their internal training businesses. With the use of training management systems comes the need to overhaul and standardize all training practices, equip and train all training staff in the use of the system, and migrate employee-training information from current records to the new training management database—and this is just the beginning of the costs of using a total training management system that involves competency assessment and tracking features, as information needs to be added to the system at least annually for each employee. Even after a company has decided to create and manage competency profiles for its employees, the choice of whether to use a paper-based or a Web-based system for this purpose may well depend on how many employees need to be managed, and whether the variety or complexity of their jobs necessitate the tracking of a large number of competencies.

4.7 Future trends

Many organizations are presently at the stage of evaluating different types of distance learning, and the focus tends to be on which Web-based course-delivery platform best meets the needs of the organization. As more courses are delivered via the Web, an increasing number of companies will find that a training management system is required to ensure that employees have easy access to online courses. Web-based courses make it much easier to track whether students have taken courses and the results of online testing. Online courses even make it possible to see when employees took courses, where they left off, and which test questions they found the most difficult. During the next few years, many organizations will adopt the Web as a core mode of training delivery, and training management systems will be the next area where interest will be concentrated.

To a great extent the increased use of competency assessment and training management systems will simply be the result of the availability of these systems in the marketplace. The need to manage competencies and training is not a new idea. Back in the early 1990s the Canadian Department of National Defence purchased a custom-made competency assessment and training management system because of the high need that organization had for managing training related to the many occupational trades in the military. As similar systems become available off-the-shelf, organizations without the monetary resources to commission custom software will also be able to take advantage of the benefits of competency and training management systems.

With time, competency and training management systems will become even more sophisticated in features and functionality. Better links to human resources management systems such as PeopleSoft will allow for more financial analysis on training and performance. As well, analytic tools will likely be incorporated that can evaluate ROI and the impact of training on job performance and corporate profits. The result will be a more accurate picture of organizational performance and job performance, and perhaps closer ties between training, job performance, and financial compensation.

4.8 Conclusion

Enterprise-wide systems function as a user interface to both internal and external online training delivery systems and manage user and course information for online and classroom training. As a user interface, these systems allow employees to go directly from individual training profiles that contain management recommendations for Web-based courses, to a virtual learning center where competency assessment testing is done and the courses themselves may be accessed. The system keeps an online record for each employee and data residing in broader human resources systems such as PeopleSoft can also be linked to the training management system.

For investments in training to be most effective, information management is key. Training is one element of the performance management of individual employees and the organization as a whole. Competency and training management systems allow for the sophisticated management of information related to training. Systems such as SIGAL effectively manage information related to individual and organizational competencies, training delivery, financial management of training, and performance tracking.

References

[1] Whalen, Tammy, and David Wright, "Distance Training in the Virtual Workplace," chapter 5 in *The Virtual Workplace*, Harrisburg, PA: Idea Group Publishing, 1998.

[2] Dove, Rick, "A Knowledge Management Framework," *Automotive Manufacturing and Production*, Vol. 110, No. 1, January 1998, pp. 18–20.

[3] Zemke, Ron, and Susan Zemke, "Putting Competencies To Work," *Training*, Vol. 36, No. 1, January 1999, pp. 70–76.

[4] Rubino, John A., "A Guide to Successfully Managing Employee Performance: Linking Performance Management, Reward Systems, and Management Training," *Employment Relations Today*, Vol. 24, No. 2, Summer 1997, pp. 45–53.

[5] Malcolm, Stanley E., "Where EPSS Will Go From Here," *Training*, Vol. 35, No. 3, March 1998, pp. 64–69.

[6] Brown, Duncan, and Michael Armstrong, "Terms of Enrichment," *People Management*, Vol. 3, No. 18, September 11, 1997, pp. 36–38.

[7] Crew, Rudy, "Creating a Performance-Driven System," *Economic Policy Review*, Vol. 4, No. 1, March 1998, pp. 7–9.

[8] Davis, Steven R, Jay H. Lucas, and Donald R. Marcotte, "GM Links Better Leaders to Better Business," *Workforce*, Vol. 77, No. 4, April 1998, pp. 62–68.

[9] Gordon, Jack, "Outsourcing to the Web," *Training*, Vol. 35, No. 6, June 1998, pp. 98–99.

[10] Stamps, David, "Enterprise Training: This Changes Everything," *Training*, Vol. 36, No. 1, January 1999, pp. 40–48.

[11] Phillips, Jack J., and Patti F. Pulliam, "The Seven Key Challenges Facing Training and Development," *Journal of Lending and Credit Risk Management*, Vol. 80, No. 4, December 1997, pp. 25–30.

[12] Kaeter, Margaret, "The Automatic Training Tracker," *Training*, Vol. 36, No. 5, May 1999, ET22–ET23.

[13] Jones, John W., *Virtual HR: Human Resources Management in the Information Age*, Menlo Park, CA: Crisp Publications, 1998.

[14] Hall, Brandon, *Online Training Management Software: How To Choose a System Your Company Can Live With*, Sunnyvale, CA: Brandon Hall Resources, 1998.

[15] Stamps, David, "Enterprise Training: This Changes Everything," *Training*, Vol. 36, No. 1, January 1999, pp. 40–48.

5

Pricing Models for Web-Based Training

5.1 Introduction

In most medium and large companies, recognition of the relationship between job performance and training is demonstrated by a significant investment in employee training programs. Typically, business units will be given annual training budgets, usually based on a fixed amount per employee, to be spent on courses chosen at their discretion. In some cases, companies will determine that specific training is required for a segment of the employee population and will set aside additional funds for special training programs. In many companies, an internal transfer of funds occurs when an employee takes a course. The business unit where the employee works pays the company's training department tuition for each course taken. End users also have the option of taking courses from external vendors, so courses offered by the training department must be competitive in terms of price as well as selection and perceived quality.

From more effective and convenient training to cost savings, Web-based training has enough potential advantages that many training departments are incorporating some Web-based courses into their traditional classroom-based curriculum. Once companies have chosen whether to buy Web-based courses off the shelf or have them custom developed, as well as worked out the logistics of offering Web-based courses to employees, the question of what to charge for this new type of course naturally arises.

The pricing of goods and services is based on market demand, competition, and cost of production. This chapter examines how those forces affect the pricing of Web-based courses. We review prices currently charged by vendors of Web-based courses, describe several pricing scenarios, and discuss how they compare in terms of competitive pricing, profitability, payback period, ROI, and net present value (NPV).

5.2 Market forces

Companies are making increasing use of distance learning for employee training. Organizations surveyed in 1998 reported that only 70% of the courses delivered to employees are now delivered by a live instructor in the classroom, compared with 81% in 1987 [1]. CD-ROMs and corporate intranets are the most commonly used delivery mediums. Interestingly, 42% of organizations using distance learning report that trainees interacted with an instructor or other students during the courses. In other words, few students are left to their studies without available human support.

While many organizations are beginning to use Web-based training, this technology is still at what Geoffrey Moore [2] terms the early adoption stage. Early adopters are typically technology-minded consumers who are willing to take some risks by being among the first to use a promising new product. However, mainstream consumers, who are more price-sensitive and who ultimately decide whether a new product will make it in the long term, are not yet purchasing Web-based courses in large numbers. The companies that are the potential consumers of Web-based courses may be uncertain whether this technology has sufficient value to be worth the trouble and initial expense.

The availability of substitutes for Web-based training is high, especially in the form of classroom-based training. While ROI, as discussed below, is potentially high for companies using Web-based training, both an initial investment and planning for Web-based training implementation are required. Web-based courses need to be purchased from third-party developers or custom developed, and employee computers and networks may need to be upgraded for course delivery. Implementing a Web-based training program requires technological expertise, planning, and project management. This initial investment in both time and money may discourage the majority of companies and keep Web-based training from reaching the mainstream market.

In 1997 the market was enthusiastic about the future of training companies, with stocks selling for 30–50 times expected earnings [3]. The purchase of training companies by famed investor Warren Buffet and junk bond legend Michael Milken had investors paying close attention to the industry. However, vendors of online courses had a rough time meeting the expectations of their shareholders in 1998. Even the stock value of CBT Systems, now Smartforce, the market leader in online courses, plummeted in the fall of 1998 due to an unexpected failure to meet revenue targets, and over the following six months only slowly regained an upward trend [4]. Indications are that the demand for online courses may not be as high as is required to support even the limited number of vendors currently competing in the marketplace.

The willingness of vendors to increase the supply of Web-based courses is primarily limited to expectations of consumer demand and their ability to invest the $45,000–135,000 per course hour necessary for development costs [5]. Course production prices in Canada for asynchronous courses are in the $35,000–65,000 (Canadian) range. Additional funds are also required for product marketing. The larger online course vendors offer consumers hundreds of titles, and unless a new entrant concentrates on a very specific niche market, investment could be substantial and the risks of not recovering costs high. While barriers to entering the Web-based course market are certainly not nearly as high as starting a new business in which large capital costs must be incurred, substantial investment is still required. As with any new technology-based product, gaining acceptance by mainstream customers poses real challenges.

Companies entering the market as producers of Web-based courses must also have technical staff familiar with off-the-shelf multimedia software products and access to expertise for authoring content and for instructional design. Since many companies now producing Web-based courses have a background in producing other types of multimedia-based products—and, in fact, sometimes continue that line of business concurrently with their course production businesses—entry into the market of Web-based courses is relatively easy. Another common entry strategy is to combine expertise in classroom training with complementary Web-based course offerings. Classroom training companies already have access to content experts and instructional designers and have a potential market for Web-based courses in their current classroom customers. In terms of pricing, multimedia and classroom training companies can keep costs relatively low, as distance learning is an extension of their current business practices and only incremental investment must be made. For both multimedia and classroom training companies, exit from the distance-learning market is also not difficult, as their former product lines are still salable.

Searching the Web for the availability of online courses and examining the list of course producers in *Training* [6] indicates that there are a larger number of vendors for IT training than for other types of courses such as soft skills training, general interest, or industry-specific courses. Several vendors offer technology-based courses for high-demand subjects such as SAP. For example, Smartforce/CBT Systems and NETg are among several vendors for this type of training. Although a large number of people will need SAP training within the next few years, there is a predictable end to the demand for SAP courses that will occur when the market for SAP is saturated and fewer companies are installing the system.

Considering that online course vendors appear to be having some difficulty making sales, the market may be in a situation of oversupply with respect to current demand. To make a reasonable profit and still keep prices to the consumer low, Web-based course producers need to sell courses to more than one customer rather than develop custom courses that can be sold only once. This may explain why there seems to be relatively few manufacturing-related courses available in the marketplace, and a proliferation of courses designed to teach common office skills such as word processing.

5.3 Factors in competitive pricing

There is typically a range of prices for any product category that is acceptable to consumers. Monroe [7] suggests that customers will pay more for a product they associate with high quality, and their perception is often based on the following factors:

1. Perceived value of the product;
2. Intrinsic benefits of the product;
3. Relative prices of similar products;
4. Consumer knowledge of the prices of competitive products;
5. Customer expectation that current prices will not undergo drastic increases;
6. Reputation of the product vendor.

5.3.1 Perceived value

From a corporate perspective, the delivery costs for Web-based courses are significantly less expensive than those for classroom training, making the use of this technology potentially a good investment. The business case for Web-based training is normally based on the amount that is spent on in-house course production or acquisition from a third party, the number of students who will take a course, and the expenses that will be offset from reducing or eliminating classroom-based training. Results of the Bell Canada business case study described in Chapter 2, which examined the development and delivery costs for three courses, the expected number of students who would take the courses each year, and the savings from eliminating classroom delivery of the courses, indicated that Bell Canada should invest in Web-based training for its employees.

As discussed previously, ROI is the financial gain or loss that results from making an investment. For Bell, the ROI was high for all of the courses, indicating that there was value in Web-based training for the company.

Another consideration is the comparative value of one particular Web-based course over another, although this is a difficult judgment to make until a consumer has taken courses from several vendors to compare the various products. With exposure to a variety of Web-based courses, consumers will be able to comparatively assess factors such as depth of course content, clarity of presentation, how engagingly content is presented, and the level of interactivity and ease of use. The difficulty in assessing quality and value of Web-based courses has resulted in the creation of new businesses in which consultants do quality assessments for Web-based courses on the market. This service allows buyers to have some idea of the value of the product before purchase. Alternatively, some vendors such as Computer Literacy are attempting to assist consumers by estimating the number of hours it takes to complete the courses they sell so potential buyers have an idea of the substance of the materials presented in the course [8]. Another solution is for vendors to provide consumers with minicourse samples to allow them to see what the full course would be like. However, since the purpose of a course is to acquire knowledge, to truly gauge the effectiveness of a course, precourse and postcourse testing needs to be done.

The fundamental question arises of just what constitutes a "good course." Factors that play a role include the following.

- The attributes of the product, such as the quality of the content;
- The quality of presentation, such as the quality of the instructional design;
- Desired features, such as a high level of multimedia elements or characteristics such as simulations using virtual reality.

A greater use of multimedia increases the cost of Web-based course production. However, it is not clear that more multimedia elements add value by increasing retention of the course materials. If the value of multimedia is aesthetic, what is the right balance between course attractiveness and price? Fister [9] believes that sound instructional design principles such as planning content around the needs of students, presenting relevant information in an engaging manner, and focusing material on course objectives are far more important than including expensive multimedia elements to present information. Obviously, if the price tag for creating a course is minimal, the number of students that need to take a course to break even through cost savings over classroom training is less of an issue. The concern is whether low-cost courses are engaging enough for effective learning and for students to complete the course. The price charged for such a low-cost course could range from free to quite high if the information conveyed in the material is in high demand and difficult to obtain from alternative sources.

5.3.2 Intrinsic benefits

Some of the intrinsic benefits of Web-based courses are their ability to allow students to take the course from anywhere with a computer and network connection, do the course at any time, and proceed at the student's own pace, reviewing course materials as required. The perceived value of Web-based training to potential students will depend on whether they believe that these benefits outweigh the advantages of classroom-delivered courses. In favor of classroom training are the advantages of human interaction, both with instructors and peers; no requirement for additional learning in the use of computers and networks before taking the course; and the advantage of familiarity.

Another example of an intrinsic benefit of a Web-based course would be one offered by a college, university, or accrediting organization that leads to a diploma, degree, or other certification. Many colleges and universities offer programs that can be partially or entirely completed over the Web. As well, certification programs such as the Microsoft Certification series may also be completed through Web-based courses.

5.3.3 Relative prices

The relative prices of similar products include alternatives such as classroom training, technology-based training using other modes of delivery such as CD-ROM, and, of course, the prices of other Web-based courses on the same topic. For example, classroom courses at Bell Canada are priced at approximately $300 per day, while companies such as Learning Tree charge approximately $750. Since Web technology is new, vendors are charging a wide range of prices for Web-based courses and alternatives. For example, the training firm 7th Level is providing 150 online multimedia courses to the 12 million subscribers of America Online for prices ranging from $19.95 for a six-month subscription to $200 for Microsoft certification courses. By comparison, CD-ROMs sell for about $40 [10].

Courses that teach "commodity skills" required by a large number of people, such as word processing, tend to command far lower prices than courses for specialized knowledge, such as SAP training. Some commodity courses are selling for as little as $4.95 per hour, with typical courses being sold for $30 to $40. At the other end of the spectrum, some technical certification courses are priced close to $500 per course [11].

5.3.4 Consumer knowledge

Consumer knowledge of classroom course prices is high for training departments and at least moderate for student end users. Training departments are expert buyers of training products and familiar with pricing for all types of courses. Most potential students have experienced classroom-based training, and a good number have also had the opportunity to use CD-ROM course materials.

The prices of competing Web-based courses are less well known, as this technology is still fairly new. However, the level of knowledge a training department would have of Web-based course pricing would likely be moderate, while for potential students, the level of knowledge would be low. As professional buyers of training products, training departments would tend to research the prices for Web-based course alternatives, at least before making the decision to purchase a

Web-based course. Potential students may be less likely to undertake the time-consuming task of exhaustively researching alternative Web-based courses to compare prices.

5.3.5 Customer expectation

Consumers normally expect that prices of products will either increase gradually over time or fluctuate in either direction with advances in technology. In the case of new products such as Web-based courses, the initial price set in the marketplace by the initial vendors of the product will influence consumers' expectations of the level of pricing for the product. If vendors set prices very low to tempt customers to try the product and expect that they can significantly increase prices as the product becomes more popular, they are likely to find that use of the product drops off drastically after prices are raised [12].

Vendors of new products must weigh whether charging a higher initial price that results in slower market penetration outweighs the risk of setting a price without a sufficient profit margin. The risk of setting an initial price too low is that producers may never be able to convince consumers that a higher price is the fair market value for their product. Buyers are far more likely to be sensitive to price increases than price decreases, meaning that sales can easily be lost from increasing a price but not necessarily gained from decreasing a price [13].

5.3.6 Vendor reputation

The reputation of a course vendor is complicated by the fact that the area is new, and most consumers have not taken enough Web-based courses to be highly aware of the quality of the courses offered by the various vendors. As well, consumers usually must take the risk of buying a course sight unseen, which is not how we typically buy other products. For example, when customers purchase material goods such as a new laptop computer, the customer can see, touch, and try the computer before purchasing. The purchaser likely has a clear idea of the different features available on new laptop machines. Consumers of goods can also make clear judgments about whether particular options are worth the price considering both need and budget and likely have a good idea of what to expect from the offerings of the different manufacturers. Computers are commonly available at a store where the buyer can try several computers and easily compare differences in the quality of such features as screen display, usability, and speed of processing. By comparison, purchasing an online course, particularly when the purchaser is the end user rather than a professional in a training department, leaves the customer with many uncertainties.

Although data about consumer behavior when it comes to the purchase of online courses is not available, some vendors of online courses have built a strong presence in the industry from long experience. For instance, many potential customers, especially professional trainers, recognize the name of Smartforce/CBT Systems, and its products are therefore more likely to be regarded as high quality even among those who have not actually seen the company's products. In the cases where the course vendor is a university, the reputation of the university itself will serve as an indication of course quality to the potential buyer.

5.3.7 Price elasticity

Elasticity of demand is the degree to which consumers are sensitive to increases and decreases in the price of a product. In general, consumers are most sensitive to prices that are relatively high or low compared with the average price of similar products [14]. However, this principle applies only to products that are priced within a range that is relatively close to the average price.

Products at the extremes of the pricing scale are less elastic than those priced closer to the average. The reason for this is that quality is a key factor in consumers' willingness to pay more or less for a product. Consumers of premium products place a high value on the product's quality and are willing to pay almost any price to get the quality they desire. At the other extreme, consumers of inferior goods cannot afford higher priced products and will tend to purchase low-cost items regardless of relative quality. As Winkler [15] points out, the price itself is often the consumers' most significant indicator of quality. Extending this concept, price reductions would be expected to result in an increase in sales of products where the consumer expects the quality of all brands to be uniform. However, when the consumer equates low price with low quality, a decrease in price is likely to result in a decrease in sales for that brand.

Another consideration is whether a product has unique attributes or unique benefits [16]. Consumers will normally pay more for products for a special purpose than those more closely associated with day-to-day needs. Carliner [17] points out that Web-based courses that are mission-critical and contain proprietary content will be priced higher than courses that are available from several sources and contain commodity-type content. In other words, consumers would be expected to pay more for courses required for a special project—for example, SAP implementation—than for courses on topics such as word processing, which are used by many people on a regular basis.

Other factors that make consumers more sensitive to price are listed as follows.

- The relative dollar magnitude of the purchase (i.e., buyers are more sensitive to the price of big-ticket purchases);
- The frequency of past price increases (i.e., if the price of a product has recently increased, buyers may not have adjusted to the last price change and will be unwilling to accept further increases);
- Practical barriers to consumption, such as time or money constraints, or lack of need.

Another factor that increases elasticity of demand is a large number of substitutes. As Devinney [18] points out, prices will be more sensitive when a greater number of substitutes are available, but the lack of substitutes does not guarantee lower price sensitivity. In the context of Web-based training, a large number of producers of courses on a particular topic should indicate that prices are generally favorable to the consumer. However, a lack of competition may simply mean that there is little or no demand for Web-based courses on a particular topic.

5.4 Elements of course pricing

When Web-based course vendors set prices for their products, there are typically six elements that must be considered:

1. Course production costs consisting primarily of labor and equipment;
2. Costs of obtaining intellectual property (i.e., course content from a third party);
3. Overhead costs for items such as server hosting, software licenses, bandwidth, help desk, administration, management, and real estate;
4. Fair profit margin on costs;
5. Comparative prices of similar courses available in the marketplace (i.e., substitutes);
6. Consistency with product positioning and corporate image.

5.4.1 Course production

The business case in Chapter 2 established the production costs of Web-based training. Table 5.1 lists the average costs (in Canadian dollars) the study found for the various elements of asynchronous Web-based courses.

Brandon Hall's research on the costs of producing asynchronous Web-based training courses, reported in *Inside Technology Training* [19], indicates that 150–300 development hours are required for each training hour of a Web-based course, depending on the amount of multimedia elements such as graphics, audio, and video that are included. Per-hour production charges are roughly equivalent to per-hour charges for general multimedia production. For a typical Web-based course, costs range from $45,000–135,000 per finished hour. With an exchange rate on the Canadian dollar of 1.45, this translates into $65,250 to $195,750 (Canadian) per finished hour. By comparison, the two asynchronous Web-based courses produced for the Bell Online Institute, each lasting 2.5 hours, cost $34,680 and $63,445 (Canadian) per training hour to produce.

Hall also reports that the learning curve and reuse of material will account for about a 50% reduction in the development time to produce a second course over the first one and that improvements in development time will continue to occur with the production of subsequent

Table 5.1

Average Costs for Various Elements of Asynchronous Web-Based Courses

Production Function	Cost (in Canadian dollars)
Instructional design	$26,877
Course authoring	$9,171
Text production	$4,041
Multimedia design	$11,849
Graphics production	$20,748
Photo production	$953
Audio production	$2,860
Content integration	$12,848
Training	$4,086
Tests	$6,038
Testing integration	$3,133
Modification/adjustment integration	$5,993

courses. In economic terms, Hall illustrates how the incremental, or marginal, cost of producing one more unit (i.e., one more course) decreases with mass production. In other words, the Web-based training industry, like other industries, benefits from economies of scale. For a company that produces Web-based courses to set an appropriate price, the company must consider its average cost of production as the baseline for calculating prices for a typical course.

5.4.2 Intellectual property and overhead

Costs for intellectual property will be incurred when elements of course content are acquired from another party (for example, SAP course content from SAP Inc.). A third type of cost is overhead cost normally incurred by the business unit that produced, or contracted for the production of, the course. This would include, for example, management costs and the real estate costs of housing employees.

5.4.3 Fair profit margin

Fair margin on the production of the course is a fourth type of cost. Although it might be argued that companies would benefit if training departments charged business units for courses at cost, modern accounting practices suggest that the internal transfer of goods and services between business units is best done at fair market value [20]. In other words, business units should compensate training departments at market rates for the courses they offer.

Likewise, business units should have the option of buying from external course suppliers if the training department cannot provide satisfactory goods and services at a reasonable price. The reason for this is twofold. First, if internal suppliers were compensated at less than a fair market value, they would not be able to reinvest profits to improve their product offerings. Training departments that are viewed as cost centers are vulnerable to cutbacks whenever companies decide that they have to control costs. When many areas of a company have to compete for scarce resources, the valuable role of training is likely to be forgotten. However, in the long run, companies will spend large sums on external training suppliers, which could likely be better spent within a training department that tailors course offerings to a company's unique needs. On the other hand, forbidding business units to purchase courses from external sources could lead to a decline in the quality of in-house offerings due to the lack of competition.

5.4.4 Product positioning and corporate image

Product positioning and corporate image are pricing considerations that are related to understanding the needs and mind-set of the consumer. Positioning takes into account both the value proposition of the product and the company's corporate image. Value proposition refers to the features that differentiate one product from that of a competitor including flexibility, speed, and price. Corporate image is similarly tied to concepts—conveyed by terms such as premium provider, value provider, and low-cost provider—that differentiate one company from another. These concepts are essential for training departments as well as Web-based course vendors to properly align courses with customer needs and successfully promote the courses that are being offered. Product price needs to match product positioning and corporate image to maximize the effectiveness of the marketing effort and consumer demand.

5.5 Pricing models

Rebecca Ganzel [21] points out that neither vendors nor trainers have a clear idea of what price is reasonable for Web-based courses. Different vendors are using different pricing models, resulting in vastly different prices for courses. For example, universities, for the most part, charge either the same price as a similar classroom course for their Web-based offerings or charge more using the logic that a distance course offers students the added value of convenience. Ganzel defines four common models currently used to price Web-based courses:

1. Charge per course hour;
2. Charge by monthly subscription;
3. Charge per fixed period (e.g., 10 weeks) for self-study courses;
4. Charge per element of course content (e.g., a video clip that can be included in a database).

Carliner [22] further defines the "per course" model by specifying the following.

- Same price as classroom, using the logic that a Web-based course can be reused at no additional charge, plus have more convenience;
- Higher price than classroom, common among universities, which use the logic that students will pay more for the convenience of taking a course at a time and/or location more convenient for them than the regular classroom.

Carliner also adds a fifth pricing model called "value-added." In this model, Web-based courses come bundled with another product that has been purchased—for example, a software package—and is used to train the purchaser in the product's use. A twist on this scenario is when a manufacturer of a product makes Web-based training available free to resellers of their product, for the reason that knowledge of the product will increase reseller sales.

In some cases, delivery method determines pricing model. Web-based courses tend to charge by the hour or by a monthly subscription, while CD-ROMs are priced by the course. On the other hand, Digital Education Systems sets its prices for Web-based courses at one-quarter to one-fifth of the price for an equivalent classroom course. The company believes that students and trainers will be more willing to use their product to supplement classroom training if the price is right [23]. The company also believes that the vast number of potential customers on the Internet will more than make up for a lower price in the volume of product sold.

Another company, DigitalThink, prices its Web-based courses the same as equivalent classroom courses. Courses vary in price depending on whether the subject area is common (e.g., word processing) or is less usual in the marketplace (e.g., SAP). In other words, a course in high demand but with many suppliers will command a lower price than a course with lower demand and fewer suppliers. When market forces are applied, the price that is charged is what the customer is willing to pay.

Traditional classroom courses were priced according to how much time was spent in the classroom. However, Web-based courses can be self-paced, where the concept of course length is elastic. Furthermore, there is evidence that a Web-based course can convey a similar amount of "learning" in less time than an equivalent classroom-based course. This is often referred to as course compression, and rates reported in the literature typically range from 20% to 80% [24]. For example, if a company believes that employee training will result in fewer errors on the job and a value can be placed on those errors, those particular courses could be priced just below the value of the training to the company. The difficulty, of course, lies in assessing how much students are able to learn from any particular course (i.e., the value of the course to the student) and how to compare the merits of one course to another before buying (i.e., course quality).

5.6 Pricing scenarios

From the reports on prices vendors are charging for Web-based courses, a midrange cost for production and overhead is about $130,000 (Canadian) for a two-and-a-half-day course based on the courses analyzed for Bell Canada in Chapter 2. The first scenario in Table 5.2 illustrates a cost-plus pricing model. A profit margin is added to the known development and overhead costs and then divided by the expected number of students per year to arrive at a price per student with a payback period of one year.

The NPV is the difference between an investment's market value (or savings on current operations—i.e., traditional classroom-based training) and the project's costs. Described as an equation, NPV would be equal to:

$$\text{Sum of savings on current operations}/(1 + k)^t - \text{Total costs}/(1 + k)^t$$

The discount rate, or cost of capital (k), is the minimum required return on a new investment to break even on the project's capital costs. In other words, the discount rate represents the opportunity costs of making a capital investment, and the chosen rate must be at least equal to the going rate for market investments with the same level of risk. Present value $(1 + k)^t$ is the current value of future cash flows dis-

Table 5.2
Scenario 1

Model	Cost	Margin	#Students	Payback	Price
Cost-plus	$130,000	10%	170	1 year	$850
(Altered		20%			$956
margins)		30%			$1,093

counted at the appropriate discount rate. In Tables 5.3, 5.4, and 5.5, 12% was used as the discount rate.

ROI is the percentage that represents the net gain or loss of using Web-based training instead of classroom delivery. As an equation, ROI can be stated as:

$$(\text{Present value of total savings}/\text{Total fixed costs}) \times 100$$

Using the above prices, the NPV of the project and the ROI calculated over a three-year period are summarized in Table 5.3.

Although both the NPV and ROI are attractive, based on evidence reported in the literature that a median price in the marketplace is approximately $250, a cost-plus model with payback in the first year results in relatively high prices. By fixing the margin at 20% and varying the payback period, a different pricing scenario emerges. The resulting unit prices are shown in Table 5.4.

Under this scenario, the NPV and ROI are still attractive. Using the above prices, the NPV of the project and the ROI are calculated over a three-year period in Table 5.5.

Clearly, even lower prices could be charged if either the initial production cost for the course were lower or if the number of students who enrolled in the course each year were higher. While no vendor is able to charge consumers prices that will not recover costs within a reasonable amount of time, one would still expect a range of prices for similar courses in the marketplace. Some vendors and products will be positioned as premium brands, perhaps because the vendor offers a wide selection of courses with high-quality content. Depending on how the vendor and courses are positioned, prices for individual courses will tend to be higher or lower than the average.

5.7 Case study: Pricing considerations for Web-based courses at Bell Nexxia

Bell Nexxia is Canada's largest provider of broadband and internetworking telecommunications products and services. The company

Table 5.3
Project NPV and ROI

Price	NPV	ROI (%)
$850	$216,896.49	266.84
$956	$260,297.58	300.23
$1,093	$316,188.19	343.22

Table 5.4
Unit Prices

Model	Cost	Margin	Number of Students	Payback	Price
Cost-plus (Altered payback)	$130,000	20%	170	1 year	$956
				2 years	$478
				3 years	$319

Table 5.5
Three-Year NPV and ROI at 20% Margin

Price	NPV	ROI (%)
$956	$260,297.58	300.23
$478	$65,148.79	150.12
$319	$99.19	100.08

is involved in several Web-based training initiatives, both to upgrade the technical skills of employees and to resell a Web-based service, Nexxia.IP Learning, to Nexxia customers. While Bell has developed some Web-based courses in-house, the majority of its course offerings come from external vendors. Bell Nexxia hosts the courses on its own servers, markets and delivers the courses to both internal and external customers, and technically produces course materials for the Web as required.

The primary considerations for pricing Web-based courses at Bell Nexxia (Chris Gill, personal communication, April 14, 1999) are:

- The costs of developing a course, including instructional design, subject matter expertise for course authoring, and production, plus a profit margin;
- The total number of "transactions" in a given year, based on the number of courses offered times the expected number of students who will enroll in those courses;
- The transaction costs, which include a portion of management, marketing, software, equipment, bandwidth, maintenance, and real estate overhead plus a profit margin on those costs associated with hosting the courses on Bell Nexxia servers;
- The prices of competing courses in the marketplace.

Bell Nexxia expects to launch 17 60-minute self-paced technology courses, 17 15-minute sales-on-the-run Nexxia service courses, 36 1-hour expert-online instructor-led courses, and 36 Nexxia Service Online courses in the second quarter of 1999. There are currently no plans to restrict the amount of time a student may spend on a course. The initial market for these courses is Bell Nexxia's 300 salespeople. Other potential customers for the courses are the thousands of sales staff at other BCE companies that include Bell Canada and Nortel, as well as Bell's corporate customers. While prices are still being finalized, prices for the courses will be in line with the prices of similar courses in the marketplace. For example, higher prices will be charged for longer courses and for those that are on topics that are in demand but not widely available from other sources. The prices of the courses will be the same to both internal and external consumers.

5.8 Conclusion

While Web-based training is still at the early stages of the market cycle, this technology is nonetheless increasingly used for employee training. In the past 10 years, the number of companies using some form of technology-based training has increased from roughly 20% to 30%. While current prices for Web-based courses vary greatly from one producer to another, pricing trends appear to be following predictable models of lower prices for "commodity" topics and higher prices for unique courses on topics in high demand. Elements that must be considered when setting prices are the costs of course development, intellectual property, profit margin, and the pricing of competitive courses in the marketplace. Evaluating various pricing scenarios indicates that factors such as production cost, projected number of students trained per year, profit margin, and expected payback period will have dramatic effects on unit prices. However, considerations that are not related to cost, such as product positioning, corporate image, and the prices being charged for competing courses, must also be taken into account. In the case of Bell Nexxia, pricing decisions by marketers are consistent with both pricing theory and the practices of other pricing experts working in the area. Prices charged to internal business units are the same as those charged to external customers of Web-based courses.

References

[1] Lakewood Publications Inc., "Training Magazine's Industry Report, 1998," *Training,* Vol. 35, No. 10, October 1998, pp. 43–45.

[2] Moore, Geoffrey A., *Crossing the Chasm: Marketing and Selling High-Tech Products to Mainstream Customers*, New York: HarperBusiness, 1995.

[3] Bassi, Laurie J., Scott Cheney, and Mark Van Buren, "Training Industry Trends 1997," *Training and Development*, Vol. 51, No. 11, November 1997, pp. 46–59.

[4] Yahoo! Inc., *Statistics at a Glance – CBTSY: Stock Performance, May 1988–March 1999* (http://biz.yahoo.com/p/c/cbtsy.html), April 6, 1999.

[5] Hall, Brandon, "The Cost of Custom WBT," *Inside Technology Training*, July/August 1998, pp. 46–47.

[6] Lakewood Publications Inc., "Just Ask: How To Price Training Converted for the Web," *Online Learning News,* Vol. 1, No. 13, June 29, 1998.

[7] Monroe, Kent B., *Pricing: Making Profitable Decisions,* New York: McGraw-Hill, 1990.

[8] Ganzel, Rebecca, "What Price Online Learning?" *Training*, Vol. 36, No. 2, February 1999, pp. 50–54.

[9] Fister, Sarah, "Web-Based Training on a Shoestring," *Training*, Vol. 35, No. 12, December 1998, pp. 42–47.

[10] Lakewood Publications Inc., "Low Cost Training: Try It—Cautiously," *Online Learning News*, Vol. 1, No. 49, March 9, 1999.

[11] Carliner, Saul, "What To Pay, What To Charge," *Online Learning News*, Vol. 1, No. 47, February 23, 1999.

[12] Monroe, Kent B., *Pricing: Making Profitable Decisions*, New York: McGraw-Hill, 1990.

[13] Ibid.

[14] Ibid.

[15] Winkler, John, *Pricing for Results*, London: Heinemann, 1983.

[16] Monroe, Kent B., *Pricing: Making Profitable Decisions*, New York: McGraw-Hill, 1990.

[17] Carliner, Saul, "What To Pay, What To Charge," *Online Learning News*, Vol. 1, No. 47, February 23, 1999.

[18] Devinney, Timothy M., "Economic Theory and Pricing Behavior," in *Issues in Pricing: Theory and Research*, Lexington, MA: Lexington, 1988.

[19] Hall, Brandon, "The Cost of Custom WBT," *Inside Technology Training*, July/August 1998, pp. 46–47.

[20] Garrison, Ray H., George Richard Chesley, and Raymond G. Carroll, *Managerial Accounting: Concepts for Planning, Control, Decision Making*, Homewood, IL: Irwin, 1990.

[21] Ganzel, Rebecca, "What Price Online Learning?" *Training*, Vol. 36, No. 2, February 1999, pp. 50–54.

[22] Carliner, Saul, "What To Pay, What To Charge," *Online Learning News*, Vol. 1, No. 47, February 23, 1999.

[23] Ganzel, Rebecca, "What Price Online Learning?" *Training*, Vol. 36, No. 2, February 1999, pp. 50–54.

[24] Hall, Brandon, *Web-Based Training: A Cookbook*, New York: Wiley, 1997.

6

Business Process Reengineering for the Use of Distance Learning at Bell Canada[1]

6.1 Introduction

BOLI represents a radical change in the way Bell Canada provides internal training to its 35,000 employees. BOLI specializes in Web-based training, one type of technology-enabled (distance) learning. Web-based training is a significant departure from the more traditional

1. Reprinted with the permission of Idea Group Publishing, *Annals of Cases on Information Technology Applications and Management in Organizations*, Vol. 1, 1999, pp. 186–199.

classroom-based practices at the BIPD, the business unit that oversees all employee training at Bell Canada. This case study examines the use of Web-based training at Bell Canada in the context of BPR. We present a theoretical context and a practical guide to how technology-enabled learning changes the business processes in an organization. The study defines the processes that are required to deliver Web-based training, the value to the internal and external business practices of the organization, and the costs for each process. The wider applications of this case study are identified and will be of interest to those in organizations that are moving from classroom-delivered training to distance delivery.

This case study describes changes in the organization that result from reengineering, including the impact Web-based learning has on training plans, student needs assessments, the ability to provide specialized curricula, training students and instructors in using new technologies, and establishing a principle of continuous improvement. Alternative ways of achieving project objectives are presented, along with organizational impact, technology alternatives, and cost benefits.

6.2 Background

The telecommunications industry in Canada has undergone dramatic changes within the past few years,due to deregulation, rapidly changing technologies, and the globalization of business practices. A downsizing effort at Bell Canada in response to the changing nature of the telephone industry has affected almost every area of the company, with the internal training function being no exception. BIPD, responsible for all internal training activity at Bell Canada, has been reinvented over the past three years, and changes are still ongoing (Stephanie Sykes, personal communication, March 11, 1997). The first change occurred in 1995, when all training that had once been conducted in-house by BIPD was outsourced to four companies known as the "training partners." BIPD's role changed from that of training provider to strategic planner. The second change occurred in 1997, when a decision was made at the corporate level to increase the use of distance learning for training delivery to Bell Canada employees. While the process that BIPD went through to transform training at Bell Canada was not called "reengineering" within the company, BIPD carried out an extensive planning process that was then applied to the "6 R's" methodology, described below.

6.2.1 Methodology

In a work that predates the concept of BPR, Jim Stewart [1] sums up the role of training in achieving organizational change when he states that such change can happen only when individual change occurs through learning. While there are a great number of published methodologies for applying BPR concepts, one model that emphasizes organizational transformation is the 6 R's of BPR by Johnson A. Edosomwan [2].

6.2.1.1 The 6 R's

Edosomwan's 6 R's methodology draws upon previous well-known works on BPR. Michael Hammer, the originator of the BPR concept, emphasizes the need for complete redesign of the organization to accomplish reengineering objectives. In *Reengineering the Corporation*, Hammer and Champy [3] identify innovation, speed, customer service, and quality as key success factors in the process. Other authors have also applied the original concepts of BPR to specific situations such as training. In *Reengineering the Training Function*, Donald Shandler [4] describes preparation, process identification, vision development, solution formulation, and transformation as major reengineering elements and clearly shows how they apply to corporate training.

The advantage of the 6 R's approach is that the methodology is concise yet well defined. Below is a summary of the 6 R's methodology.

1. *Realization:* Realization is the understanding that dramatic changes are needed to meet the needs of organizations, teams, and individuals, and support the changes in the organization that occur due to rapidly evolving economies and technologies; and that the redesign practices must fit with the organization's business objectives.
2. *Requirements:* The identification of strategic direction for the reengineering effort requires an analysis of customer needs.

Decisions may then be made about the products and services that will fill the gap between the desired outcome and the processes, products, and services that are needed.

3. *Rethink:* New structures and systems must be put in place to accomplish the objectives of the reengineering program. In this stage, new procedures, processes, and technologies are identified.
4. *Redesign:* At the redesign stage, specific targets are set for implementing the reengineering process.
5. *Retool:* At the retooling stage, technologies or other more competitive systems are evaluated and adapted to the reengineering project.
6. *Reevaluate:* During reevaluation, performance indicators are measured to determine how well the reengineering effort met its goals.

6.3 Setting the stage

Distance learning is an example of a changing business practice that Bell Canada hopes will improve both the bottom line and the delivery of training, which is a significant benefit to employees. BOLI was created in order to explore the possibilities for using technology-enabled learning in the organization. Web servers were procured for this purpose, and a core team was established with expertise in corporate training delivery, technology, and business development. Other interested parties were consulted as necessary, such as representatives from end-user groups. Distance learning has several obvious advantages over classroom training, namely the following.

- Convenience and flexibility in the time, geographic location, and pace of training delivery;
- Cost savings over current classroom-based training based on reduced training time and travel savings.

A related change to the organization was the establishment of Bell Learning Solutions (BLS), a business unit that was formed to provide distance-learning services to customers through consulting, reselling third-party products and services, and creating and marketing technology-based course content.

6.4 Case description: 6 R's analysis

6.4.1 Realization: Establishing the need for business process reengineering

There has been a great deal of interest in distance learning within Bell Canada. One of the reasons that BOLI was established was that Bell Canada believes that technology-enabled training adds value to the training program. Employees find attending classes increasingly difficult to coordinate with their work schedules and believe that distance learning will offer them more convenience and flexibility than classroom-based training. Classroom training is an "event," requiring preregistration and courses offered at fixed periods of time, at a location apart from an employee's normal place of work. Distance training delivery to the desktop is one alternative that can provide continuous learning to employees. From a management perspective, distance learning offers cost savings over current classroom-based training and enables employees to acquire and retain the skills necessary for Bell Canada to compete effectively in a global marketplace.

6.4.1.1 Outsourcing

Before 1995 BIPD conducted all employee training at Bell Canada. BIPD had a $40 million budget, classroom space, and 426 employees working in course design and delivery, as well as in administrative support functions such as project management, technology support, registration, and course logistics. In 1995 training at Bell Canada was outsourced to four training partners. One company provides telecommunications technology-related training, a second company provides IT-related training, a third company provides soft skills training in areas such as leadership and management, and BOLI provides support for distance learning. Courses once offered by BIPD were on the topics of telecommunications, information systems and technology, and general business skills, the same subject areas now covered by the training partner companies.

After outsourcing, the BIPD budget was $9 million and staff was reduced to 60 employees. Training partners also now provide classroom space as one of their outsourcing services, along with all required course materials. The demand for employee training remains strong. In 1997, 14,000 students, equivalent to 43,300 student days, were taught at BIPD through the training partners.

6.4.1.2 Current assessment

The current situation points to several possible directions for BIPD. These include the following.

- The need to reduce cost of training delivery;
- The need to update course curriculum frequently so that topics of current interest are provided, meeting the needs of the workforce and reflecting the evolving computer and telecommunications technologies used in the organization;
- The need to provide high-quality training;
- The need to give end users flexibility in their training location;
- The need to reduce training time by focusing training very specifically on topics needed by clients to carry out their jobs;
- The need to align training with profit-making business ventures of the company (i.e., the support of BLS, the business unit that designs and delivers tailored solutions to customers interested in implementing distance learning in their own organizations).

6.4.2 Requirements: Defining customer needs

6.4.2.1 BIPD's strategic role

BIPD now functions primarily in a strategic role to set the direction for training within Bell Canada and is the liaison with the outsource training partners. Liaison activities include building business relationships, identifying pricing and contract issues, and ensuring that internal customer needs are met. A customer services team of BIPD learning professionals has specialists dedicated to specific groups of Bell Canada employees to ensure that employee training needs are addressed. For example, a specialist may do an evaluation of an employee group to identify performance gaps and then put the employees in touch with one of the training partners to help fill those gaps through appropriate courses.

There is no transfer pricing done between the business units of employees taking courses and BIPD. However, the training partners do charge Bell Canada business units for courses employees take. Business units are given their own training budgets, which they may use where they wish, and are not limited to the courses offered by training partner companies only. If an employee does take a course through one of the training partners, payment is made directly to the partner company and BIPD gets no part of the payment. Tuition fees vary from course to course, depending upon course features such as the length of the course, the venue, and whether there is a need for equipment rental.

Both supervisors and employees may make the decision to take a course at BIPD, and balancing the needs of these two client groups is important. On the one hand, employees need the opportunity for personal development; on the other, the needs of the corporation must be at the heart of BIPD's program. There is shared accountability between the employee and the employer when an employee takes a course; when an employee signs up for a course, a copy of the registration form goes back to the employee's supervisor.

The idea for a new course often comes from the training "customers" themselves. For example, a call center supervisor may feel that there are problems with his staff that a formal training session could remedy. BIPD would respond by using a set of diagnostic tools to determine whether the supervisor is correct and a course is the best solution to the problem. If the development of a new course is indicated, the next step would be for BIPD to bring in the appropriate training partner, who would use a standard instructional design systems process to move from the "idea" stage to developing an actual course. If a suitable instructor is not already on the staff of the training partner, the partner company would find an expert in the field and, if necessary, train her or him to be an effective trainer.

Before outsourcing, BIPD operated at a tactical level, performing all aspects of employee training in-house, from instructional design to course delivery. Now, BIPD spends much more time identifying how training can address the needs associated with supporting Bell Canada's key business issues—for example, changes in the technologies the company sells or even change management issues themselves. Apart from planning training strategy and management of outsourcing partnerships, BIPD is also still responsible for executive education, employee orientation, career management, technology support, and administrative support.

Included in BIPD's strategy for the coming year is the exploration of training options with the highest learning value to ensure a competitive advantage for Bell Canada. Within that strategy, the organization wants to evolve the technology-based learning capability within Bell. BIPD's key success factors in implementing this strategy are listed as follows.

- The availability of human and financial resources;
- The shared commitment of the training partners;
- The availability of suitable computer systems to end users and the ability of end users to adapt to a technology-enabled training environment;
- The ability to move quickly within a shorter cycle time;
- The availability of learning applications that meet the needs of customers and the business needs of Bell Canada.

Table 6.1 presents an analysis of the strengths, weaknesses, opportunities, and threats at BIPD.

6.4.2.2 Recommendations

From the analysis in Table 6.1, several next steps can be identified:

Table 6.1
Analysis of BIPD Before Distance Training Delivery

Strengths	Weaknesses
Outsourcing has created opportunity for flexibility in training provision Outsourcing has meant that BIPD functions on a strategic rather than an operational level Senior management is committed to using new technologies for training delivery and for new profit-making opportunities	Classroom training has high cost Course curriculum does not always reflect end-user needs for current topics of interest, due to evolving job requirements and changing technologies
Opportunities	**Threats**
End users requesting training at their desktops Needs could be met through alignment of BIPD and BLS	Deregulation of telephone industry has increased need for broader business opportunities Changing technologies and globalization have increased need for current information

- Research costs of Web-based training;
- Research Web-based learning platforms;
- Test usability of platforms and acceptance of distance training through a pilot project;
- Identify further courses for distance delivery within the coming year;
- Get commitment of training partners to support a distance learning strategy;
- Increase use of Internet/intranet delivery of training.

6.4.3 Rethink: Integration of appropriate technologies

6.4.3.1 Distance learning initiatives at Bell Canada

The challenges of deregulation, globalization, and constantly evolving technologies represent both threats and opportunities for the company. On the one hand, increased competition has necessitated drastic reorganization and painful cost cutting; the current reality induces high levels of stress in employees, who are faced with new, and often unknown, demands. On the other hand, employees are rising to the challenge and evolving the business into new areas that draw on many strengths in the company. One of the new ventures is the provision of consulting expertise and distance learning outsourcing to customers through BLS. Bell Canada has 20 years of experience in the distance education field. Bell has explored the internal use of various distance training methods and has also provided customers with distance learning technologies such as videoconferencing and business television. The many business activities in the company have allowed for the redeployment of much expertise related to distance learning, including a wide variety of Internet and electronic commerce applications.

Distance learning, like the current classroom training, is outsourced to the training partner companies. In-house, BIPD is developing expertise in instructional design related to multimedia materials in order to work effectively with the training partners. Partnership with BOLI is BIPD's first initiative in Web-based training. BOLI functions as a testing ground for various Internet-based learning platforms used for training delivery at Bell Canada.

BOLI has no direct reporting relationship with BIPD. Because BOLI has been independent of the existing training infrastructure, there has been more opportunity for creativity in that organization. BIPD uses the products that BOLI develops to provide services to the end users. Courses that are developed by, or for, BOLI become part of the BIPD curriculum, which is updated and managed by the training partners.

The use of distance learning has encountered no resistance from BIPD staff because it represents an interesting challenge rather than a threat. The real changes to training delivery at Bell Canada happened when outsourcing occurred several years before distance learning was introduced. To BIPD, distance learning is simply a service supplied by the training partners in response to "customer" demand.

Profit-making opportunities for Bell Canada have increased with the establishment of BOLI. The basic products of BLS, the business unit created to resell Bell Canada expertise in the area of distance learning to external customers, are consulting services, course production in a distance format, and reseller services for course content from many different sources. BOLI acts as the technical resource to fulfill customer needs for course production and technical expertise in consulting projects. While BIPD, BLS, BOLI, and the other training partner companies are independent organizations, they are able to work together to meet a wide variety of internal and external customer needs. For example, BOLI, BIPD, and BLS have partnered with an SAP implementation partner to design and deliver SAP training internally to Bell Canada employees.

One of the other benefits that Bell Canada expects will come with the use of distance learning is that courses will be available to employees without a formal registration and approval process. As well, employees will not need to leave their normal place of work. The result may be an increase in the amount of training delivered to employees, making employees well-equipped with knowledge that they can use to deal with constantly changing job demands, right at their fingertips.

6.4.3.2 Continuous improvement and alternative delivery options

In choosing an appropriate technology to deliver the courses in the distance learning pilot, the advantages and disadvantages of various modes of delivery were analyzed.

Web-based course delivery is one of the alternatives to classroom delivery. Other technology options include business television, videoconferencing, audiographic conferencing, and CD-ROM or other type of CBT. Depending on instructional design considerations that in turn depend on the needs of users and the type of course material being presented, these technologies could be used as alternatives to Web-based delivery or as an extension of the capabilities of the Web. For example, CD-ROMs may contain hotlinks to Internet or intranet sites, and videoconferencing and business television can broadcast Internet applications to students.

A major aspect of BPR is continuous improvement. In Bell Canada's case, this means continuously evaluating which technologies are most appropriate for the training required. Tables 6.2 and 6.3 give a comparison of the advantages and disadvantages of distance delivery using CD-ROM, intranet, Internet, audiographic or videoconferencing, and business television, as compared with classroom delivery. Comparisons are made on the basis of organizational impact and costs.

6.4.4 Redesign: Setting goals for performance improvement

BIPD established a 1998 goal of moving 20% of the curriculum to a Web-based distance learning platform. This represented a 4% increase in the materials available in a distance learning format, and is equal to an additional 32 courses. Most of the 32 courses will be developed and delivered as Web-based training, as BIPD feels that offering courses via the Bell intranet is a solution that best meets the needs of most employees.

The Web offers significant advantages over CD-ROM-based training, such as the ability to update materials quickly and inexpensively and the ability to reach virtually all employees without the need to physically distribute course materials. There is also the benefit of the larger storage capacity of a server as opposed to a CD-ROM, allowing the use of more multimedia elements such as animation and video. The primary means of distance delivery is expected to be through the corporate intranet or the Internet. Due to relatively high costs and limited flexibility, little use of videoconferencing or business television is expected.

BIPD presently has 150 independent study courses on CD-ROM or diskettes. It is assumed that all of these courses could potentially be taught using any of the asynchronous platforms. There are presently 700 classroom-based courses offered through BIPD. It is estimated that 10 percent of these could be potentially offered using the Centra Symposium synchronous platform.

In order to judge the cost-effectiveness of a distance training solution, economic analysis is required and a positive return on investment

Table 6.2

Comparison of Classroom to Distance Delivery Using CD-ROM, Intranet, or Internet

	Classroom	CD-ROM	Intranet (Web)	Internet (Web)
Mode	Live	Asynchronous	Synchronous/asynchronous	Synchronous/asynchronous
Availability to Bell Canada employees	High (although some employees must travel)	High	High	High
Possible methods of communication	Live lectures, Q&A	Telephone Fax E-mail	E-mail Discussion groups Live text chat Synchronous audio or video	E-mail Discussion groups Live text chat Synchronous audio
Student and instructor interaction	High	Low	Medium (potential for real-time text, audio and video)	Medium (potential for real-time text and audio)
Interactive exercises	Limited (although breakout possible)	Yes	Yes	Yes
Effective course presentation	Excellent (interaction with instructor possible)	Good (limited communication)	Excellent (interaction with instructor possible)	Good (graphics, video, and real-time communication slow)
Easy access to external resources	No (although may access Internet from separate application)	Yes (if use hotlinks to connect to Internet/ intranet sites)	Yes (if have Internet connection)	Yes (to other Web sites)
Easy to use	Yes	Yes (with initial instruction)	Yes (with initial instruction)	Yes (with initial instruction)
Easy to update course information	No (if use printed course materials)	No (need to repress and redistribute)	Yes	Yes
Distribution cost	Medium/high depending on travel required	Low/medium (depending on number of updates)	Low	

Table 6.3
Comparison of Classroom to Distance Delivery Using Audiographic or Videoconferencing, or Business Television

	Classroom	Audiographic Conferencing	Video-Conferencing	Business Television
Mode	Live	Synchronous	Synchronous	Synchronous
Availability to Bell Canada employees	High (although some employees must travel)	Medium (requires multimedia PC or equipped room)	Medium (requires multimedia PC or equipped room)	Low (requires broadcast studio and receiving sites)
Possible methods of communication	Live lectures, Q&A	Live voice Whiteboard	Live audio and video Can include white board	Live audio and video Can include whiteboard
Student and instructor interaction	High	High	High	Medium (potential for real-time communication)
Interactive exercises	Limited (although breakout possible)	No (although breakout possible)	No (although breakout possible)	No (although student polling possible)
Effective course presentation	Excellent (interaction with instructor possible)	Excellent (interaction with instructor possible)	Excellent (interaction with instructor possible)	Excellent (interaction with instructor possible)
Easy access to external resources	No (although may access Internet from separate application)	No (although may access Internet from separate application)	No (although may access Internet from separate application)	No (although may access Internet from separate application)
Easy to use	Yes	Yes (with initial instruction)	Yes (with initial instruction)	No (requires scripting and television production)
Easy to update course information	No (if use printed course materials)	No (if use printed course materials)	No (if use printed course materials)	No (if use printed course materials)
Distribution cost	Medium/high depending on travel required	Medium	High	High

is determined. To test the economic benefits and feasibility of using Web-based training in the organization, a pilot project was designed.

6.4.5 Retool: Testing of the concept

6.4.5.1 Bell Online Institute pilot project

As discussed in Chapter 2, BOLI piloted three Web-based courses in 1997 to measure the cost and evaluate the effectiveness of training delivered on four different Web-based learning platforms.

Before these three Web-based courses were developed, 16% of BIPD's curriculum could be offered in a distance format, but these course materials were on paper or audiovideo cassettes. The pilot courses were delivered to engineers working in Bell Canada's Advanced Communication Systems group.

The course on the topic of routing was delivered on the Centra Symposium platform. The pilot course was delivered to 20 Bell Canada engineers from Quebec City, Montreal, Ottawa, Toronto, and Hamilton. The course was offered in two parts, one in the morning and one in the afternoon. The morning session lasted two and a half hours and the afternoon session lasted one and a half hours, for a total of four hours. The engineers were present for the entire course.

The course was highly successful, with participants reporting that they felt their needs were being met with flexible, convenient course delivery on topics that were of current interest for their job activities. Testing indicated that learning for all courses was as effective as classroom training.

6.4.6 Reevaluate: Evaluating the new process

6.4.6.1 Course production costs

Cost is one of the key performance measures in BPR. In Chapter 2, we examined the costs for the business process needed to produce a Web-based course at Bell Canada. These processes include: instructional design, multimedia design, the production of text, audio, graphics, photographic and video elements, authoring and software development, content integration and modification, and training and testing design and delivery.

Figure 6.1 compares each costing element across all courses. There was some variation in the use of any particular element among the courses, although the variations were small, and some elements were consistently used more or less frequently for all of the courses. None of the costing elements could be discounted as unimportant, and none of the costing elements clearly dominated the others in importance.

The cost for each of these processes varies with the number of development hours required. As can be seen from Figure 6.1, there were some variations in multimedia elements used and in the length of development time for each. To get a better picture of costs for the development of each element, an average of the time spent on developing each element and an average of the hourly rates charged by the three production companies were used and results summarized in Figure 6.2.

6.4.6.2 Economic Return

Like all large corporations, Bell Canada routinely analyzes new business opportunities to determine economic attractiveness. However, this study is the first to calculate the economic return on an internal training initiative. Using classroom course delivery costs at BIPD as a baseline, the cost savings of using Web-based training were analyzed. Course development costs for Web-based training are higher than those for classroom courses, but savings result from lower delivery costs. After calculating the savings in Web-based delivery per student, the ROI was determined as well as the number of students that must take the course before the high development costs were recovered through savings in course delivery over the client-server architecture. The case study analyzed the break-even number of students required to recover Web-based course development costs and the ROI over a five-year period. All of the measures of financial performance indicated that the business case for Web-based training is strong.

The break-even point of an investment project is the point at which either cost savings or revenue generated equals the cost of investment. Realizing savings for Web-based courses requires a sufficient number of students to recover course development costs. In this case, the break-even point is the number of students that must be trained for the

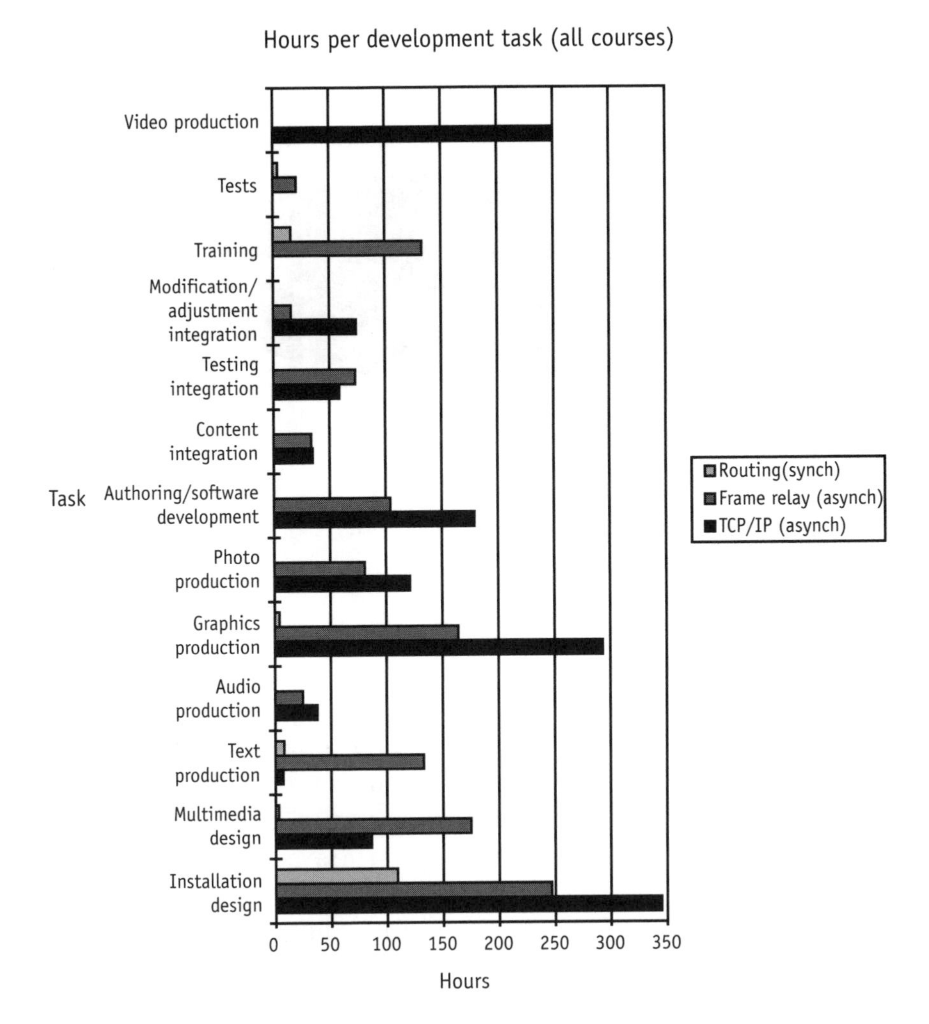

Figure 6.1 Hours per development task (all courses).

fixed costs of Web-based course development to be offset by the reduced delivery costs of Web-based training. The break-even points for the three Web-based courses in this pilot ranged from four to 112 students.

ROI is the financial gain (or loss) that results from an investment project. ROI was positive for all of the Web-based courses in the study.

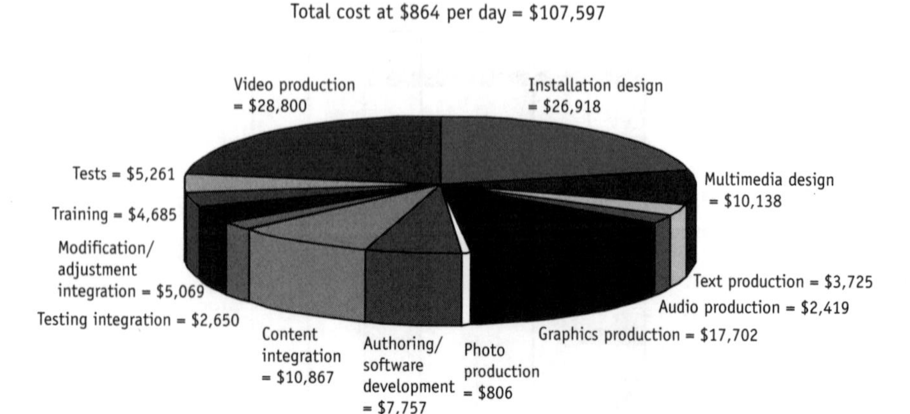

Figure 6.2 All courses: average cost per developer task. Total cost at $864 per day: $107,597.

ROI ranged from $3 saved for every $1 spent on Web-based training to $33 saved for every $1 spent. The average savings per student was $702 for the asynchronous courses and $1,103 for the synchronous course. Of the asynchronous courses, the TCP/IP course using the Mentys learning platform had the least cost savings, at an average of $625 per student. The asynchronous frame relay course saved an average of $850 per student.

The most critical factor in the cost benefit of using Web-based training was found to be the amount of multimedia content in a course. Courses with a higher amount of multimedia content, such as the TCP/IP course, which contained a five-minute video segment, showed a correspondingly lower ROI. This was the only course that used video and, as a result, the number of hours spent in multimedia production was far higher than for the other courses. The synchronous course, which was the most cost-effective course in the pilot, contained a few graphics and live audio only. The limited amount of multimedia content in the synchronous course offset the higher costs of course delivery that resulted from having a live instructor present during delivery, as well as greater student salary costs due to the extra hours required to deliver the course. ROIs for the Bell Canada courses ranged from $33 saved for each $1 spent on Web-based training for the course with the least amount of multimedia content to $3 saved for every $1 spent for the course with the most multimedia content.

6.4.6.3 Training quality

BIPD is involved in the continuous improvement of training quality. To accomplish this continuous improvement, training effectiveness must have a means of being measured. Kirkpatrick's [5] four levels of training evaluation, an industry standard for evaluating training, is the system currently used at BIPD.

The evaluation of training quality is ongoing in the organization. Course effectiveness is measured at Level 1 evaluation, through student surveys about their satisfaction with the instructor, course materials, and venue. In 2000, BIPD intends to move to Level 3 evaluations, in order to determine postcourse effectiveness. Participants, supervisors, peers, and possibly customers will be surveyed to evaluate whether an employee's training has made a difference to his or her performance in the workforce. Also in the near future, BIPD will use a Level 4 type of evaluation, where ROI will be determined for selected courses. This Level 4 evaluation will be done because of demand from senior management at Bell Canada to be informed about training effectiveness.

Table 6.4

Summary of Multimedia, Break-Even, and ROI

Course	Multimedia Development Hours	Break-Even Number of Students	ROI Over 5 Years
Routing	144 hours	4 students	3,283 percent
Frame relay	1,156 hours	51 students	697 percent
TCP/IP on WebCT	1,487 hours	111 students	288 percent
TCP/IP on Mentys	1,487 hours	112 students	283 percent

Table 6.5
Levels of Training Evaluation

Level 1	Reaction evaluation: how well participants liked the training
Level 2	Learning evaluation: principles, facts, and techniques that were understood and absorbed by the participants
Level 3	Transfer-of-learning evaluation: transfer of training skills to knowledge on the job
Level 4 (BIPD Goal)	Results evaluation: impact of training on the organization

A substantial amount of the expertise that was available within BIPD before outsourcing is still available to the organization. In fact, some former employees of BIPD now work for the training partner companies; others found new jobs within Bell Canada. Quality control is an ongoing issue in the outsourcing relationship. BIPD works very closely with the training partners to ensure that standards are maintained. The use of Level 3 evaluation techniques may help reduce some of the need for this close relationship.

The curriculum has improved since outsourcing. Outsourcing was the impetus needed to review the existing curriculum, and many courses were found to be outdated. In some cases, courses were updated, in others the courses were dropped from the curriculum. As well, some courses did not have detailed written documentation, and outsourcing was an opportunity to ensure that documentation was brought up to date.

6.5 Current challenges

Today, BOLI is steadily expanding the number of Web-based courses available to employees and has also been working very closely with Technomedia Training, a company that is developing a competency and training management system for Web-based course delivery. A record for each Bell Canada employee has been created, and current efforts are focused on developing suggested online training curricula based on each employee's job. Links will also be made to the enterprise-wide computer system used to manage human resources, in order to track courses employees have completed. BOLI is involved in the delivery of customer projects generated by the sales and marketing staff of BLS as well. The number of permanent staff assigned to BOLI has been increased, and there has also been an increase in the number of outside consultants brought in to work on customer projects. The current challenge to BOLI is to remain focused on projects that will add the greatest value to the organization, as well as to effectively manage the many projects that are now under way.

6.6 Conclusion

The BOLI pilot project is an effort that illustrates some general principles for reengineering classroom training to distance learning:

- The need for BPR was defined at the realization stage.
- Customer requirements were established in the requirements phase and both employees and senior management supported the proposed changes.
- Appropriate technologies were integrated into the new business process in the rethink phase.
- The reengineering project had clear goals for performance improvement set in the redesign stage.
- The change was carefully managed at the retool stage by carrying out a pilot project to test the feasibility of the solution.
- There was an evaluation of the new process based on clear performance measurements during reevaluation.

Table 6.6 summarizes the 6 R's, Bell Canada's reengineering of training, along with generalization to other reengineering efforts in the area of corporate training. The 6 R's methodology was effectively applied to the case of distance training at Bell Canada and could be generally applied to BPR efforts for training in other organizations.

Table 6.6

Application of 6 R's to Bell Canada and General Applicability to Distance Learning

Step in Process	Aim	Bell Canada	Generalization to Other BPR Effort
1. Realization	Understanding of the need for reengineering effort to better meet the needs of the organization	Classroom training not meeting the needs of employees or the company Management and employee interest in distance learning Outsourcing has created opportunity to try new methods	Review current training strategy to evaluate opportunities for improvement Evaluate opportunities for change using internal and external resources
2. Requirements	Identification of strategic direction so BPR effort meets customer needs	Identification of employee and organizational training requirements Identification of appropriate training partner to deliver specific courses Increase use of Web-based training	Review employee and management training goals Identify new technologies or competitive business system Identify whether best met with internal or external resources
3. Rethink	Procedures, processes, and technologies required are identified and put in place	Ensure that organization is structured to accommodate reengineering effort Draw on strengths in organization to take advantage of existing technologies and expertise already in place	Ensure that organization can meet reengineering effort required by evaluating organizational structure, existing technologies, and processes required Identify resource gaps
4. Redesign	Specific targets set for reengineering implementation	Reaffirm chosen technology Set goals for use of Web-based training	Plan reengineering implementation by setting targets for new methods of training design and delivery using specific technology
5. Retool	Technologies evaluated and adapted to reengineering project	Design and delivery Web-based training pilot using several different platforms for evaluation Collect data on time and costs for course development and delivery	Design and delivery pilot of technology-based training to test feasibility Collect data necessary to evaluate pilot

Table 6.6 (continued)

Step in Process	Aim	Bell Canada	Generalization to Other BPR Effort
6. Reevaluate	Performance indicators measured to evaluate reengineering effort	Conduct cost-benefit analysis of pilot Evaluate alternative delivery options to ensure continuous improvement	Conduct cost-benefit analysis of pilot Evaluate alternative delivery options to identify those that best meet needs of organization

References

[1] Stewart, Jim, *Managing Change Through Training and Development*, London: Kogan Page, 1991.

[2] Edosomwan, Johnson A., *Organizational Transformation and Process Reengineering*, Delray Beach, FL: St. Lucie Press, 1996.

[3] Hammer, Michael, and James Champy, *Reengineering the Corporation*, New York: Harper Business, 1993.

[4] Shandler, Donald, *Reengineering the Training Function: How To Align Training With the New Corporate Agenda*, Delray Beach, FL: St. Lucie Press, 1996.

[5] Kirkpatrick, Donald L., "Techniques for Evaluating Training Programs," *Training and Development*, Vol. 33, No. 6, 1979, pp. 78–92.

Appendix 6A: Supporting material for case study on business process reengineering at Bell Canada

1. What is the overall problem presented in this case study?

Traditional classroom training for Bell Canada employees was expensive and lacked flexibility when it came to delivering courses to any geographic location, at any time, and in a self-paced mode. A BPR process was required to move the organization from traditional training methods to technology-based training methods.

2. What are the factors affecting the problem(s) related to this case?

Training delivery had been outsourced for a number of years before the use of distance learning was explored by management.

3. What are the managerial, organizational, and technological issues and resources related to this case?

Although the BIPD, the organization that had once provided traditional training, had been reduced in size in 1995, it was still able to play a strategic role in planning and managing training delivery in the company. BOLI was created to function as a "training partner" to deliver distance learning to employees. While BIPD and BOLI are separate organizations, they work together closely to design and deliver distance courses.

BLS was established at roughly the same time to sell technology-enabled learning consulting and systems implementation to external customers. There are synergies between BIPD, BOLI, and BLS. The three organizations draw on their strengths to work together on technology-enabled learning projects both internally and to external customers.

Bell Canada employees are connected to a corporate intranet with multimedia PCs, and through the efforts of BIPD and BOLI they are able to take training from their desktops. Web-based training has been the most common mode of distance course delivery. The ROI for Web-based training was established in a study described in Chapter 2. While the cost of Web-based course development is higher than classroom course development, the decreased costs of course delivery has resulted in savings from $3 to $33 for every $1 spent on course delivery over a five-year period.

4. What role do different players (decision makers) play in the overall planning, implementation, and management of the information technology applications?

BIPD maintains contact with end-user business units to help determine their training needs. The next step is to engage one or more of the training partners to design and deliver the courses that are required. When end users need distance training, BOLI takes responsibility for the required course design and delivery. BOLI also researches the various technology alternatives and makes recommendations for the most appropriate mode of course delivery. BOLI also works on customer projects generated by BLS.

5. What are some of the emerging technologies that should be considered in solving the problem(s) related to the case?

Web-based course delivery over a corporate intranet was the method that the organization used successfully. Other technology alternatives include Web-based delivery over the Internet, audiographic or videoconferencing, business television, and CD-ROM or diskette-based course delivery.

Epilogue

Long-range effects

Bell has increased the use of Web-based training for employees and has made a further investment in a training management system to be used for managing online training internally. Bell has also made an equity investment in the company that developed the system.

Lessons learned

Major lessons learned from the case study are:

- Researching the pros and cons of technology alternatives is essential before implementing a system.
- If restructuring the organization is required, this must be done before attempting to implement the chosen technology.
- The chosen solution should be tested in a pilot project and data should be collected to make (or fail to make) a business case before investing in full rollout.

Further reading

Carr, David K., and Henry J. Johansson, *Best Practices in Reengineering: What Works and What Doesn't in the Reengineering Process,* New York: McGraw-Hill, 1995.

Guthrie, Edward R., "Belief and Behavior: Which Is the Chicken, Which Is the Egg?" *Management Review*, Vol. 84, March 1995, p. 62.

Hammer, Michael, "Reengineering Work: Don't Automate, Obliterate," *Harvard Business Review*, Vol. 104, July/August 1990, pp. 104–113.

Infor, Vol. 33, No. 4, November 1995, pp. 104–113.

Infor, Vol. 34, No. 1, February 1996, pp. 1–58.

Koonce, Richard, "The Human Face of Change," *Training and Development*, Vol. 50, January 1996, p. 23.

Conclusion

In a world of fast-paced change, employees are constantly challenged to learn more and learn more quickly in order to keep up with demanding job requirements. Web-based learning allows organizations to maximize the learning experience by efficiently distributing appropriate learning materials to employees' desktops, allowing access to these materials on demand, and by tracking the training process to ensure that it meets both organizational and individual needs. Web-based training links corporate training to both the IT and human resources departments.

Web-based training is the most recent in a series of technological interventions in the training arena, and for many reasons, from cost to management to reach, holds a great deal of promise as a mode of training delivery that will increase in popularity over time. As we have seen, Web-based training provides a significant reduction in the costs of course delivery. When a sufficient number of students are trained using

Web delivery, these savings offset the increased costs of development over classroom courses. Bell Canada's experience has shown that the cost benefits of Web-based course delivery can be considerable.

The break-even points for the three Web-based courses in the Bell Canada pilot ranged from four to 112 students and ROI ranged from $3 saved for every $1 spent on Web-based training, to $33 saved for every $1 spent. The average savings per student was $702 for the asynchronous courses and $1,103 for the synchronous course. Savings were realized from the ability to train many students without incurring large incremental costs (unlike videoconferencing, for example), as well as from a reduction in the number of hours required to deliver a course over the Web. The latter was due to the ability of students to skip sections they already understood as well as tighter instructional design of the course material.

Industry analysts estimate that only about 50% of the hours for an equivalent classroom course is needed to deliver a course over the Web [1]. At Bell Canada, the most critical factor in the cost-benefit analysis of using Web-based training was found to be the amount of multimedia content included in courses. Courses with a larger amount of multimedia content, such as video, had a higher number of production hours that accounted for higher production costs.

Many organizations have taken a cautious approach to implementing Web-based training, and this is hardly surprising considering that Web-based learning requires a substantially higher investment for course development than required for classroom-based learning. This is why case studies such as the Bell Canada study here are important to demonstrate how Web-based training is able to provide significant benefits for real-life corporations and to illustrate how other organizations can apply the same principles to evaluate the potential benefits of Web-based training in their own situations.

Web-based training has also created new opportunities to automate the management of the training function. The inclusion of training administration features is one of the ways that vendors differentiate their Web-based training delivery systems in the marketplace. Competency and training management systems allow job-based competency profiles, appropriate Web-based courses, and student tracking to be integrated with great effectiveness. These systems help focus training activities on real customer needs. They also act as a broadcast medium to communicate organizational training plans to employees, link individual performance evaluations to training plans, provide the vehicle for delivering the Web-based courses recommended right to employees' desktops, and keep track of student progress and success. Training management systems offer either some or all of the following features:

- Online competency assessment tools for employee evaluations;
- Searchable database of organizational competencies;
- Employee profiles that track requirements for specific courses;
- Online course registration;
- Integrated access to Web-based courses;
- Online testing and student tracking;
- Online management of training budgets;
- Links to human resources management systems such as PeopleSoft.

The more closely training is linked to job requirements, the more effective it will be. Web-based training management systems allow information about the training process to be collected, managed, and analyzed, thereby increasing the effectiveness of the training program. Trainers involved in online training delivery are starting to think of training materials in terms of "nodules of knowledge," pieces of information that support job requirements that are both less than a course and less than a course module. Trainers are also coming to realize that even technology-based learning doesn't need to happen in isolation. Groups of three to five learners can often very effectively share computer equipment and work through learning material together [2].

An important issue for organizations interested in adopting Web-based training is the business process to get from the classroom to the Web. While most organizations using Web-based training continue to also offer courses in a classroom, even implementing Web-based training in a limited form means that many business issues must first be addressed. For example, access to computer equipment and networks, the choice of hardware and software for the training solution, administration and maintenance of the technology selected, the development of job and competency profiles, and the acquisition of off-the-shelf or custom-developed course content are just some of the many issues that must be considered. Pricing Web-based courses is another business decision that organizations face, and many factors need to be considered, such as costs of the intellectual property associated with the course content, profit margin, and competition in the marketplace. Clearly, each organization will have somewhat different questions, and solutions will differ with each circumstance. This book has attempted to provide a window into the solutions that some organizations have created and offer learning from their experience as a starting point for other organizations that want to implement Web-based training.

In fact, the least important part of the Web-based learning package is the technology itself. Clearly, technology changes and improves over time, and the costs associated with using technology decreases. The most important technology question is the choice of Web-based technology over other, proprietary, technologies. An open, standards-based, technology such as TCP/IP allows organizations to implement Web-based training using a wide selection of widely available equipment, telecommunications connections, and software packages.

Interactivity and collaborative learning features are important, both for student-instructor interactions and for student-student interactions. Interactivity can be either real-time (synchronous) or at times chosen by each individual (asynchronous). Technology will continue to improve in terms of capacity, speed, and performance. Consumers of Web-based training clearly want to have the ability to interact with others, and as technologies improve, systems that allow users to choose synchronous or asynchronous interactions, using text, graphics, audio, and video will be in demand, from Web conferencing to Webcasting.

Web-based technologies are an exciting development in the training field, and the innovations that are rapidly developing will open up even more possibilities for the application of Web-based training in many organizations. Although the percentage of total training budgets that is spent on Web-based training is still very small, many organizations that have tried Web-based training in a limited way are planning to increase their use of this medium in the coming years. While there are challenges to implementing this technology, benefits on the whole appear to outweigh the disadvantages. In the future, Web-based training is likely to become very widespread—and the delivery mode of choice for training participants.

References

[1] Hall, Brandon, "The Cost of Custom WBT," *Inside Technology Training*, July/August 1998, pp. 46–47.

[2] Lakewood Publications Inc., "'Wow' Moments at the Show," *Online Learning News*, Vol. 1, No. 26, September 29, 1998.

Glossary

asynchronous communication Network communication that is not done in real time (for example, e-mail or the use of HTML pages).

audiographic conferencing The use of real-time audio and graphics-only (using a device such as a whiteboard) among groups of people, often in specially equipped rooms.

break-even point The number of students trained whereby the reduced delivery costs of Web-based training per student offset the fixed costs of Web-based course development.

Break-even = (Fixed costs of Web-based course − Fixed costs of classroom course)/ (Delivery cost per student for classroom course − Delivery cost per student for Web-based course).

browser A computer program, such as Netscape, that allows users to access and view documents on the Web.

CD-ROM (compact disc read-only memory) A plastic aluminum-coated disk that stores information digitally as nonreflective pits or bubbles. A 5.25 in CD-ROM is capable of holding about 650 MB of information.

client An end-user computer using a browser program (such as Netscape) to receive multimedia documents from a Web server.

discount rate, or cost of capital (*k*) The minimum required return on a new investment to break even on the project's capital costs. In other words, the discount rate represents the opportunity costs of making a capital investment, and the chosen rate must be at least equal to the going rate for market investments with the same level of risk.

HTML (hypertext markup language) The standard document format used on the Web.

Internet An international network linking millions of computers via telephone lines using the TCP/IP protocol.

intranet An internal network linking corporate computers using the TCP/IP protocol.

multimedia Any combination of text, graphics, images, audio, and video that has been digitally encoded.

net present value (NPV) The difference between an investment's market value (or savings on current operations—i.e., traditional classroom-based training) and the project's costs.

$$\text{NPV} = \text{Sum of savings on current operations}/(1 + k)^t - \text{Total costs}/(1 + k)^t$$

present value $(1 + k)^t$ The current value of future cash flows discounted at the appropriate discount rate. The present value interest factor is calculated using the formula:

$$\frac{1-\frac{1}{(1+k)^t}}{k}$$

return on investment (ROI) The incremental savings (or losses) resulting from the use of Web-based training.

ROI = (Present value of total savings/Total fixed costs) × 100

server An Internet host computer using a program (such as NCSA's HTTPD) to store and transmit multimedia documents for an information service (such as the World Wide Web).

synchronous communication Real-time network communication such as videoconferencing.

TCP/IP (transmission control protocol/internet protocol) A protocol that enables universal interconnection and addressing of all Internet hosts.

videoconferencing Real-time audio and video between groups, often in specially equipped rooms.

Web A network of multimedia documents stored on Internet or intranet servers, which can be seen by using a browser. Documents are formatted with the HTML standard and transmitted with the HTTP standard.

whiteboard A shared electronic drawing board that can be displayed on each workstation.

Suggested Reading

Bates, A.W., *Technology, Open Learning and Distance Education,* New York: Routledge, 1995.

Brooks, David W., *Web-Teaching: A Guide to Designing Interactive Teaching for the World Wide Web,* New York: Plenum, 1997.

Burge, Elizabeth, and Judith Roberts, *Classrooms With a Difference: Facilitating Learning on the Information Highway,* Montreal: Chenelière/ McGraw-Hill, 1998.

Chute, Alan G., Melody Thompson, and Burton Hancock, *The McGraw-Hill Handbook of Distance Learning,* New York: McGraw-Hill, 1998.

Cyrs, Thomas E., *Teaching and Learning at a Distance: What It Takes To Effectively Design, Deliver, and Evaluate Programs,* San Francisco, CA: Jossey-Bass, 1997.

Daniel, John S., *Mega-Universities and Knowledge Media: Technology Strategies for Higher Education,* London: Kogan Page, 1998.

Driscoll, Margaret, *Web-Based Training: Using Technology To Design Adult Learning Experiences,* San Francisco, CA: Jossey-Bass, 1998.

Hall, Brandon, *Web-Based Training Cookbook,* New York: Wiley, 1997.

Harasim, Linda, et al., *Learning Networks: A Field Guide to Teaching and Learning Online,* Cambridge, MA: MIT Press, 1995.

Harrison, Nigel, *How To Design Self-Directed and Distance Learning: A Guide for Creators of Web-Based Training, Computer-Based Training, and Self-Study Materials,* New York: McGraw-Hill, 1998.

Haughey, Margaret, and Terry Anderson, *Networked Learning: The Pedagogy of the Internet,* Montreal: Chenelière/McGraw-Hill, 1998.

Haynes, Cynthia, and Jan Rune Holmevik, eds., *High Wired: On the Design, Use, and Theory of Educational MOOs,* Ann Arbor, MI: University of Michigan Press, 1998.

Heide, Ann, and Linda Stilborne, *The Teacher's Complete and Easy Guide to the Internet,* 2d ed., New York: Teachers College Press, 1999.

Khan, Badrul Huda, ed., *Web-Based Instruction,* Englewood Cliffs, NJ: Educational Technology, 1997.

Kouki, Rafa, and David Wright, *Tele-Learning via the Internet,* Hershey, PA: Idea Group, 1988.

Kovacs, Diane K., *The Internet Trainer's Guide*, New York: Wiley, 1997.

Lasarenko, Jane, *Wired for Learning,* Indianapolis, IN: Que, 1997.

Laurillard, Diana, *Rethinking University Teaching: A Framework for the Effective Use of Educational Technology,* New York: Routledge, 1993.

McCormack, Colin, and David Jones, *Building a Web-Based Education System,* New York: Wiley Computer, 1997.

McDonald, Donald, *Audio and Audiographic Learning: The Cornerstone of the Information Highway,* Montreal: Chenelière/McGraw-Hill, 1998.

Minoli, Daniel, *Distance Learning Technology and Applications,* Norwood, MA: Artech House, 1996.

Moore, Michael G., and Greg Kearsley, *Distance Education: A Systems View,* Belmont, CA: Wadsworth, 1996.

Porter, Lynnette R., *Creating the Virtual Classroom: Distance Learning With the Internet,* New York: Wiley, 1997.

Roberts, Judith, *Compressed Video Learning: Creating Active Learners,* Montreal: Chenelière/McGraw-Hill, 1998.

Robin, Bernard, Elissa Keeler, and Robert Miller, *Educators Guide to the Web,* New York: MIS, 1997.

Schreiber, Deborah A., and Zane L. Berge, eds., *Distance Training: How Innovative Organizations Are Using Technology To Maximize Learning and Meet Business Objectives,* San Francisco, CA: Jossey-Bass, 1998.

Serim, Ferdi, and Melissa Koch, *NetLearning: Why Teachers Use the Internet,* Sebastopol, CA: Songline Studios and O'Reilly, 1996.

Skomars, Nancy, *Educating With the Internet: Using Net Resources at School and Home,* Rockland, MA: Charles River Media, 1999.

Verduin, John R., and Thomas A. Clark, *Distance Education: The Foundations of Effective Practice,* San Francisco, CA: Jossey-Bass, 1991.

Williams, Bard, *The Internet for Teachers (for Dummies),* 2d ed., Foster City, CA: IDG, 1999.

Epilogue

Training is a $60 billion juggernaut hauling corporate America into the 21st century. When classroom chalk was replaced by transparencies and then by computer projectors, the juggernaut changed direction, but only slightly, since trainers and trainees still met and exchanged the same words of wisdom in the same classrooms. Web-based training, WBT, with its greater flexibility in time, place, and instructional design, brings about a more significant change of course and even a change of gear, for not only does it allow people to be trained in a different way, but it also allows them to be trained faster and at lower cost.

We must be careful not to let Big Brother get his hands on WBT. When computerized cash registers were first introduced in supermarkets in New York City, the cashiers went on strike because the computers were monitoring how fast they hit the keys. With WBT, the computers can monitor how fast we learn, how many mistakes we make on the way, whether the learning impacted our productivity, and (maybe) whether we should get a pay raise. Whether the benefit to the shareholder outweighs the infringement of privacy, we leave a higher authority to judge. However, since we accept that our employer can monitor our email and voice mail, we may have become more complacent than those supermarket cashiers.

Juggernauts do not metamorphose overnight and, as yet, we hear few tales of classrooms echoing emptily after the students have flocked to the Web. But where will Web-based training take us in the future? Henry Ford produced not just the automobile, but also the suburbs and the way of life that goes with them. WBT could de-socialize education, as lonely students stare fixedly at screens of information pre-packaged by teachers they never meet. Ironically it could also re-socialize education as students and teachers participate in a *true web* of educational interaction of which today's Web based videoconferencing systems and chat groups are only very rudimentary precursors. Will Ford's suburbs be replaced by back-to-nature log cabins with broadband Internet access? Arguably students trained in regular classrooms are unsuited to an electronic work environment, and as electronic communication becomes the norm for business, will it also become the norm for professional training, for university education, for school, and (no surely not!) for kindergarten? After breakfast on the log cabin veranda, watching the early morning sun glisten in the dew-speckled, emerald-colored grass, will the children and adults disperse to different corners of the back yard, to study and work with others miles away via Web-enabled, voice-activated, holographic projections?

The juggernaut would have to make a pretty sharp turn to realize that future and, for those appalled by the possibility of such apocalyptic transformation, may we offer one hint of amelioration. Even in Star Trek, everyone *goes to* the Academy.

About the Authors

Tammy Whalen is a business development manager for learning technologies at Bell Nexxia, a subsidiary of Bell Canada specializing in broadband and internetworking applications. She has an M.B.A. from the University of Ottawa and has written a variety of refereed publications on the subject of Web-based training.

David Wright is a full professor at the University of Ottawa. He has provided distance education using a variety of technologies and has worked with major high-tech companies on the business case for a range of distance education alternatives. He has written over 25 fully refereed publications in a range of journals, plus many conference papers. He is the author of the book *Broadband: Business Services, Technologies, and Strategic Impact* published by Artech House, and coauthor of the book *Tele-Learning via the Internet* published by Idea Group. He is cited in *Who's Who in the World* and *Who's Who in Science and Technology*.

Index

Recent Titles in the Artech House Telecommunications Library

Vinton G. Cerf, Senior Series Editor

Access Networks: Technology and V5 Interfacing, Alex Gillespie

Achieving Global Information Networking, Eve L. Varma, Thierry Stephant, et al.

Advanced High-Frequency Radio Communications, Eric E. Johnson, Robert I. Desourdis, Jr., et al.

Asynchronous Transfer Mode Networks: Performance Issues, Second Edition, Raif O. Onvural

ATM Switches, Edwin R. Coover

ATM Switching Systems, Thomas M. Chen and Stephen S. Liu

Broadband Network Analysis and Design, Daniel Minoli

Broadband Networking: ATM, SDH, and SONET, Mike Sexton and Andy Reid

Broadband Telecommunications Technology, Second Edition, Byeong Lee, Minho Kang, and Jonghee Lee

The Business Case for Web-Based Training, Tammy Whalen and David Wright

Communication and Computing for Distributed Multimedia Systems, Guojun Lu

Communications Technology Guide for Business, Richard Downey, Seán Boland, and Phillip Walsh

Community Networks: Lessons from Blacksburg, Virginia, Second Edition, Andrew M. Cohill and Andrea Kavanaugh, editors

Component-Based Network System Engineering, Mark Norris, Rob Davis, and Alan Pengelly

Computer Telephony Integration, Second Edition, Rob Walters

Desktop Encyclopedia of the Internet, Nathan J. Muller

Digital Modulation Techniques, Fuqin Xiong

E-Commerce Systems Architecture and Applications, Wasim E. Rajput

Enterprise Networking: Fractional T1 to SONET, Frame Relay to BISDN, Daniel Minoli

Error-Control Block Codes for Communications Engineers, L. H. Charles Lee

FAX: Facsimile Technology and Systems, Third Edition, Kenneth R. McConnell, Dennis Bodson, and Stephen Urban

Guide to ATM Systems and Technology, Mohammad A. Rahman

A Guide to the TCP/IP Protocol Suite, Floyd Wilder

Information Superhighways Revisited: The Economics of Multimedia, Bruce Egan

Internet E-mail: Protocols, Standards, and Implementation, Lawrence Hughes

Introduction to Telecommunications Network Engineering, Tarmo Anttalainen

Introduction to Telephones and Telephone Systems, Third Edition, A. Michael Noll

IP Convergence: The Next Revolution in Telecommunications, Nathan J. Muller

The Law and Regulation of Telecommunications Carriers, Henk Brands and Evan T. Leo

Marketing Telecommunications Services: New Approaches for a Changing Environment, Karen G. Strouse

Multimedia Communications Networks: Technologies and Services, Mallikarjun Tatipamula and Bhumip Khashnabish, editors

Packet Video: Modeling and Signal Processing, Naohisa Ohta

Performance Evaluation of Communication Networks, Gary N. Higginbottom

Practical Guide for Implementing Secure Intranets and Extranets, Kaustubh M. Phaltankar

Practical Multiservice LANs: ATM and RF Broadband, Ernest O. Tunmann

Protocol Management in Computer Networking, Philippe Byrnes

Pulse Code Modulation Systems Design, William N. Waggener

Service Level Management for Enterprise Networks, Lundy Lewis

SNMP-Based ATM Network Management, Heng Pan

Strategic Management in Telecommunications, James K. Shaw

Successful Business Strategies Using Telecommunications Services, Martin F. Bartholomew

Telecommunications Department Management, Robert A. Gable

Telecommunications Deregulation, James Shaw

Telephone Switching Systems, Richard A. Thompson

Understanding Modern Telecommunications and the Information Superhighway, John G. Nellist and Elliott M. Gilbert

Understanding Networking Technology: Concepts, Terms, and Trends, Second Edition, Mark Norris

Videoconferencing and Videotelephony: Technology and Standards, Second Edition, Richard Schaphorst

Visual Telephony, Edward A. Daly and Kathleen J. Hansell

Wide-Area Data Network Performance Engineering, Robert G. Cole and Ravi Ramaswamy

Winning Telco Customers Using Marketing Databases, Rob Mattison

World-Class Telecommunications Service Development, Ellen P. Ward

For further information on these and other Artech House titles, including previously considered out-of-print books now available through our In-Print-Forever® (IPF®) program, contact:

Artech House
685 Canton Street
Norwood, MA 02062
Phone: 781-769-9750
Fax: 781-769-6334
e-mail: artech@artechhouse.com

Artech House
46 Gillingham Street
London SW1V 1AH UK
Phone: +44 (0)20 7596-8750
Fax: +44 (0)20 7630-0166
e-mail: artech-uk@artechhouse.com

Find us on the World Wide Web at:
www.artechhouse.com